Wout Krispijn
Nov. 1996

Bridging the Strait

The Story of the Confederation Bridge Project

To each participant of this amazing story.

Bridging the Strait

The Story of the Confederation Bridge Project

Copthorne Macdonald

Dundurn Press
Toronto • Oxford

Editor: Bill Freeman
Designer: Sebastian Vasile
Printer: Metropole Litho inc.

Canadian Cataloguing in Publication Data

Macdonald, Copthorne
Bridging the Strait

Issued also in French under the title: D'une rive à l'autre.
Includes bibliographical references.
ISBN 1-55002-281-4

1. Confederation Bridge (N.B. and P.E.I.). I. Title.

TG27. C66M32 1997 388.1'32'09717 C97-930719-8

1 2 3 4 5 BF 01 00 99 98 97

We acknowledge the support of the **Canada Council for the Arts** for our publishing program. We also acknowledge the support of the **Ontario Arts Council** and the **Book Publishing Industry Development Program** of the **Department of Canadian Heritage.**

Printed and bound in Canada.

Printed on recycled paper. 

Dundurn Press
8 Market Street
Suite 200
Toronto, Ontario, Canada
M5E 1M6

Dundurn Press
73 Lime Walk
Headington, Oxford
England
OX3 7AD

Dundurn Press
250 Sonwil Drive
Buffalo, NY
U.S.A. 14225

CONTENTS

PRIME MINISTER · PREMIER MINISTRE

I am pleased to be a part of *Bridging the Strait*.

The opening of Confederation Bridge is a historic moment for the people of New Brunswick, Prince Edward Island, and, indeed, all Canadians. It is an engineering marvel and a symbol of the good that can be done when the Government of Canada, the provinces and the private sector come together in partnership.

Taking a project of this immense size and extraordinary complexity from the drawing board to successful completion requires vision, dedication, talent and teamwork. The men and women who devoted themselves for many years to designing and building Confederation Bridge displayed these qualities in abundance. And I congratulate them.

Now that their work is done, I invite Canadians from coast to coast, and visitors from abroad, to use Confederation Bridge as their gateway to the distinctive beauty of Prince Edward Island and New Brunswick. And I look forward to its becoming a vital commercial link between these provinces, the rest of Canada and the world.

Bridging the Strait is a colourful and informative celebration of a tremendous feat of engineering and cooperation. I invite you to take a look and see how this dream became a reality.

Jean Chrétien

Minister of Public Works and Government Services

Ministre des Travaux publics et des Services gouvernementaux

CANADA

As the Minister responsible for overseeing the development of Confederation Bridge, I am proud to have been involved with one of the most important construction projects in Canadian history.

Confederation Bridge is a historic undertaking by the Government of Canada. The 13-kilometre bridge, the longest continuous span bridge in the western hemisphere, is a symbol of Canadian unity and an example of close cooperation between the Government of Canada, the provinces and the private sector. The arrangement the Government of Canada entered into with Strait Crossing Development Inc. represents a new approach to the provision of vital transportation services and infrastructure construction in Canada.

There are many reasons to be proud of our collective accomplishments in the development of this bridge, not the least of which is the historic significance of this new bridge and its deep roots in our past.

The 1873 Terms of Union between Canada and Prince Edward Island placed a constitutional obligation on the federal government to maintain "continuous communication" between the Island and the mainland. The government of Canada, in partnership with private industry and with the support of the provinces, has delivered on this commitment with a 1990's solution to a 124 year old commitment.

For me, celebrating the successful conclusion of the pivotal construction phase of this project is not only about our past accomplishments, but also about our collective future as a nation.

The Confederation Bridge initiative is a demonstration of good government and sound fiscal and environmental responsibility. It is also promoting long-term economic development and job creation.

Today, we have a new, world-class transportation service that celebrates Canada's past, Canada's present and a confident future rich in opportunity and prosperity.

I hope you will enjoy the many images captured in this book, *Bridging the Strait*, which highlights various aspects of the history of transportation across the Northumberland Strait and the construction of the bridge. This book records an important legacy in the history of our nation.

Confederation Bridge stands as an achievement that generations of Canadians can be proud of, knowing that it was Canadian ingenuity and hard work that made it a reality.

Diane Marleau

Prince Edward Island

Office of the Premier

Île-du-Prince-Édouard

Cabinet du Premier ministre

The first to ply the waters of what we now call Northumberland Strait would have been the Aboriginal people who lived among the stately pines that forested our fair Island. Their travels would have been very "weather sensitive" as they paddled their small canoes across to "the other side" to visit friends and relatives.

With the age of sail came the Europeans. Less than two centuries ago, they established themselves along the shores of the Strait. By the 1800's, scheduled sailings providing passenger and freight services to the proliferation of ports that accommodated the booming economy of the time.

Then came steam, and for a brief period the commerce carried from the wharves of Northumberland Strait was transported in vessels powered by either sail or steam, or sometimes a combination of both. Steam and the advent of the steel-hulled vessel also saw a vast improvement in the winter connections to the mainland.

Passage in winter was difficult. Often our ports were blockaded by armadas of ice, and even the steamers couldn't always get through. When that happened, the connection with the mainland was maintained by a hardy group of mariners that comprised the ice-boat service. Passengers, mail and a small amount of freight were carried across the mostly frozen Strait in open boats that had metal runners on their hulls. These small boats were pushed and pulled across the ice and rowed through the water that occurred when the shifting tides broke open the ice.

The modern era of transportation across the difficult waters of Northumberland Strait came in the middle of the First World War. In 1917, the SS Prince Edward Island arrived to ferry rail cars and automobiles between New Brunswick, and a newly created port on Prince Edward Island, named after the Prime Minister of the day, Sir Robert Borden.

This fine publication commemorates the latest development in transportation across the Strait. Just as the SS Prince Edward Island saw the end of the ice boats and the beginning of an age of convenience and continuity that Islanders had not then experienced, so shall the opening of the Confederation Bridge launch Islanders into another new era. I am confident that Islanders will, as they have in the past, grab hold of, and adapt this new era to their betterment, and to the betterment of our great nation.

Sincerely,

Pat Binns
Premier of Prince Edward Island

THE CONFEDERATION BRIDGE

On behalf of the Government of New Brunswick, I am pleased to join with the citizens of our province in commemorating the completion and official opening of the historic Confederation Bridge.

This engineering marvel is the epitome of modern technology. We are proud of the many New Brunswick professionals who joined a world-class team to make this project a showcase of engineering and construction expertise.

The Confederation Bridge is a bold and magnificent structure that not only physically joins New Brunswick and Prince Edward Island, but also symbolically links our historical and cultural roots. Since our earliest steps on the path towards Confederation, New Brunswick and Prince Edward Island have shared a vision of a unified Canada. The Confederation Bridge is one more symbol of the progress we continue to make in uniting this great country of ours.

We believe the Confederation Bridge will be a generator of economic activity and a catalyst for continuing change in Atlantic Canada. In the tourism industry alone, the bridge will be a crucial stimulus. While the tourism industry expects a permanent increase in visitors to the region of an estimated 25 percent, the bridge itself is expected to draw thousands of interested visitors each year from all over the world.

We can all be justifiably proud of the Confederation Bridge. The vision of the earliest proponents, three-quarters of a century ago, of a physical link between New Brunswick and Prince Edward Island has finally be realized.

The Confederation Bridge is the fulfillment of a dream that has abided for generations. It is entirely fitting that this dream has become a reality as we enter the new millennium.

Frank McKenna
Premier

STRAIT CROSSING

A JOINT VENTURE OF
Strait Crossing Inc.
G.T.M.I. (Canada) Inc.
Ballast Nedam Canada Limited

Strait Crossing is proud to have successfully completed the construction of one of the great bridges of the world, the Confederation Bridge. We are both honoured and pleased to participate in the commemoration and official opening of this great Canadian success story and landmark.

The successful completion of the Confederation Bridge signals the beginning of a new era for Canada. This historic facility brings Prince Edward Island and the remainder of Atlantic Canada closer together, thus placing the region into a more competitive position in all global markets.

Future undertakings around the world will be measured against the excellence and success of this project. The partnerships created between the public and private sectors as well as all stakeholders in the community at large will provide a model for the financing, designing, building, operating and maintaining of future infrastructure projects.

The partnership formed between Strait Crossing and the Government of Canada ensured that the Confederation Bridge will reduce the long term cost to the Canadian taxpayer and reduce the long term cost to the user. The Strait Crossing solution also ensured that the project would not harm the environment and would be completed on schedule for the Official Opening Ceremony on Saturday, May 31, 1997.

Strait Crossing invites all Canadians to participate in this celebration of Canadian achievement and unity. We are looking forward to the continuation of our partnerships and look forward to being part of the community as the operator of the Confederation Bridge for the next 35 years.

Sincerely,

Paul Giannelia
Project Director

PREFACE

This book is about the Confederation Bridge and how it came to be. It tells the story of one of the great infrastructure projects of this century, and I feel honoured to have had the opportunity to tell it. My modus operandi as a writer has always been to immerse myself in something that interested me, come to some degree of understanding, and then try to share both the interest and the understanding through writing about it. In this instance my immersion involved talking with dozens of people associated with the project; reading seemingly numberless books, articles, documents and reports; delving into UPEI Robertson Library's 18 bound volumes of newspaper articles on the subject; and seeing a good deal of the bridge project for myself. It has been a fascinating immersion, and I came out of it with much to tell.

The telling has been greatly enhanced by selections from a rich pool of visual materials. From public and private collections of historical materials came photos, news clippings, letters, and relevant artifacts. From the developer, designer, financier, builder, owner and operator, Strait Crossing, came technical diagrams, computer graphics, scale models, and stunning photographs — many of which were taken at Strait Crossing's request by Summerside, P.E.I. photographers Alain, Marlene and Buffie Boily to document the project. Every month or so from 1994 to 1997, the Boilys went up in their airplane with Hasselblad in hand and dozens of rolls of film. Other striking and informative photos came from Public Works and Government Services Canada, and a few other sources.

My challenge has been to capture the story of the Confederation Bridge as completely and effectively as possible in the space available. Innovative engineering, and novel, award-winning construction methods were a key part of the tale, but there was much, much more. With so rich and complex a story, finding the right balance between depth and breadth, pro and con, technical and human, was a constant challenge. If I opted for more detail here, it meant less detail there. I had to make thousands of judgment calls, and I take full responsibility for the choices made. In the end, I achieved a balance that suited me. I hope it suits you, too.

One editorial issue that needs at least passing mention is my handling of Imperial and Metric

measurements. As a nation, we still have a foot in each domain, and both appear in this book. Context was my guide. The bridge documentation uses Metric measurements, so in talking about the bridge I did the same. I've used Imperial measurements in the historical chapters and in a few other places: boat horsepower, for instance, and farm acreage.

I want to express my appreciation to the many people who were willing to talk with me about the project. Excerpts from some of these conversations appear in the book, but every conversation helped me to better understand the project and the events surrounding it. I also appreciate the support I received from Strait Crossing, especially the continual helpfulness of Krista Jenkins, and the many supportive activities of Bill Freeman, Carl Brand and Sebastian Vasile of Dundurn Press. Thanks to Harry Holman of the Provincial Archives, Nichola Cleaveland of P.E.I.'s Government Services Library, the reference staff of UPEI's Robertson Library, and Ed MacDonald of the P.E.I. Museum and Heritage Foundation. Thanks also to Lena Baker, Jim Feltham, Hubert Jacquin, and Sandra MacPhee of Public Works and Government Services Canada, and to Ron MacDougall of PWGSC who took some superb ground-level photos of project activities. Finally, special thanks to friends Deirdre Kessler, John DeGrace, and Beverly Mills Stetson for their important contributions, both to the book and to my state of mind when things got especially hectic.

Copthorne Macdonald
Charlottetown, P.E.I.

The Confederation Bridge looking from New Brunswick to Prince Edward Island. [Boily]

CHAPTER ONE

PROJECT AND PLACE

This is the story of the 12.9 kilometre Confederation Bridge which spans the Northumberland Strait in Atlantic Canada. Completed in 1997, it connects Canada's smallest province — Prince Edward Island — with the mainland. It is one of the longest continuous multi-span bridges in the world, and the longest bridge over ice-covered water.

The story of the bridge is not a single story, but a matrix of stories and substories. At one level it is the tale of a concrete and steel structure. At another, it is the story of an historical problem: the need for reliable and convenient transportation across a sometimes menacing body of water. It is also the story of a vision — the vision of a long-term solution to that problem, and the challenge of maintaining the vision until it became a reality. It is a story about smarter and wiser ways of doing things — about tailoring solutions to circumstances — about marrying design and construction activities more intimately — about creating significant public works in times of fiscal restraint — about innovative financing mechanisms — about mitigating negative socio-economic impacts — and about environmental concern, not as the developer's enemy, but embraced as a legitimate part of the development process at every stage, and with a monitoring aspect to mitigate environmental impacts and verify environmental assessment predictions.

It is also the story of more than a century of public controversy, ranging from concern over the adequacy of ice boats and ice-breaking steamships to the feasibility and appropriateness of a tunnel or bridge. On the YES side of P.E.I.'s 1988 fixed-link plebiscite were the 60% who voted for easier, more reliable, and ultimately less expensive transportation to the mainland. For them, a bridge or tunnel was clearly a personal and societal plus. It was something to be desired and supported. On the NO side were the 40% who, for a variety of reasons, opposed a fixed link to the mainland. Many felt that easier access to the Island would attract too many visitors, leading to uncontrolled development and serious damage to **the Island way of life**. (This phrase is defined somewhat differently by each individual who uses it, but it is a concept close to the heart of a great many Islanders.) An almost opposite concern was that visitors to the Island enjoy arriving by ferry, and that a bridge would *hurt* tourism. Some NO voters feared that a bridge would not hold up to ice conditions or would be downed by a ship collision, leaving the Island worse off than when it had a ferry system. Some simply enjoyed living on an island, and felt that a bridge would diminish the "islandness" of the place they love. Still others had concerns about employment loss, the environment, negative impacts to the fishery and agriculture — or the possibility that the project might start but never be completed (something that had happened once before, in the 1960s).

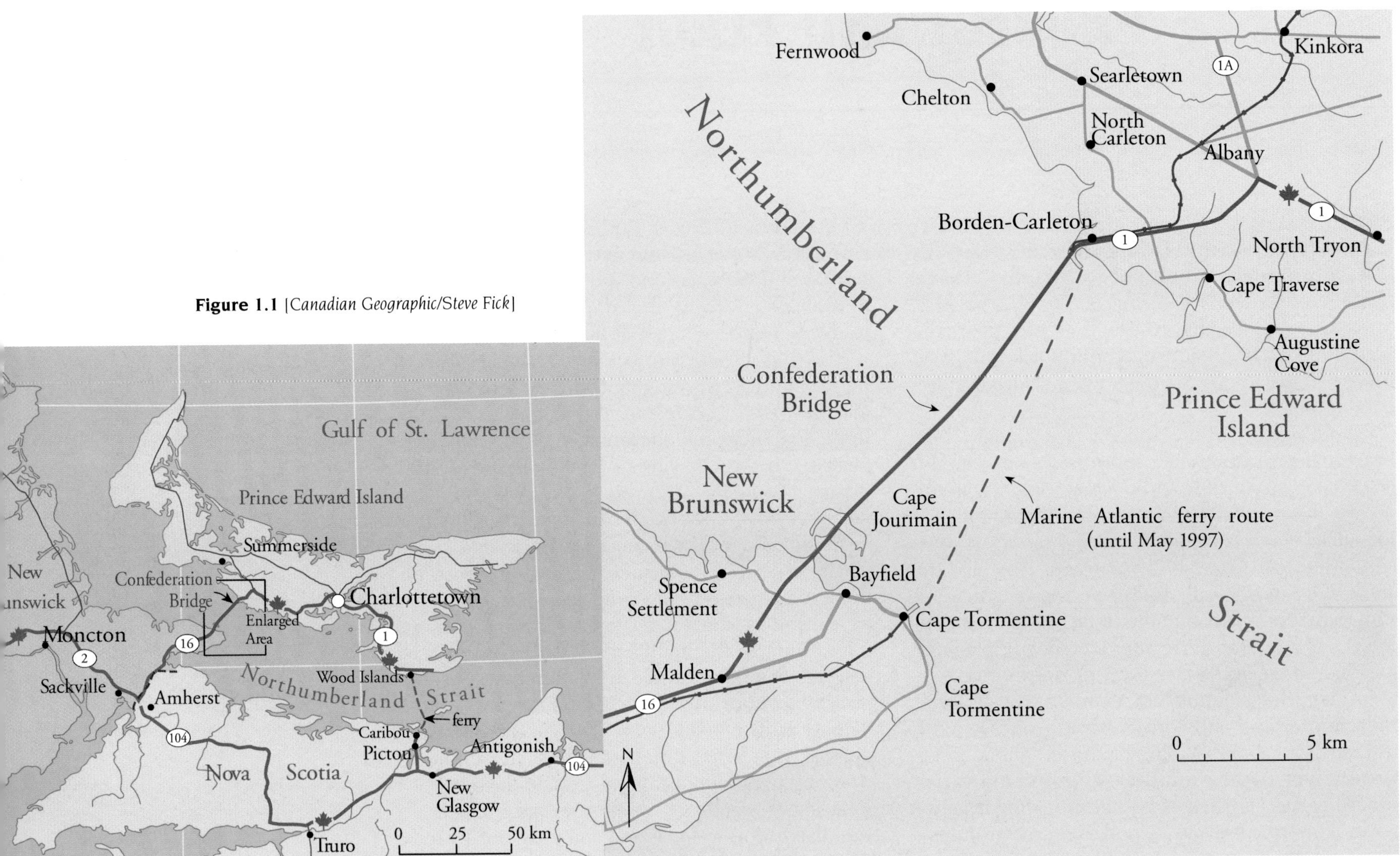

Figure 1.1 [*Canadian Geographic/Steve Fick*]

Through it all, the story of the bridge is a people story: A story of human concern and significant human accomplishment. A story of 5000+ workers and hundreds of companies from Atlantic Canada applying their energies to a task they found significant and inspiring. A story of Canadian vision and international expertise. A story of expertly coordinated, heart-felt effort.

All the substories mentioned above — and several others — will be told in the sections of the book that follow. Let's begin by looking briefly at the geographical, cultural, and economic context within which all this has happened.

Identification with Place

Most Atlantic Canadians have a strong attachment to place. Despite less than ideal weather and a scarcity of good jobs, they love their home province, the people of the region, and the way of life. For generations, economic pressures have forced young people (and older ones) to leave, but identification with their place of origin rarely disappears. Every large city in Central and Western Canada has its community of Maritimers and Newfoundlanders, many of whom would return home in a minute if they could just find a way to make a decent living back there.

Why this attachment? It seems to be linked partly to geography and partly to culture. For many there is love of the forested hills, the patchwork fields, and the sea. Atlantic Canada occupies the northeastern end of the Appalachian mountain range. Northern New Brunswick, Cape Breton, and Newfoundland are home to hills of significant size, but hills overlaid with soils that are thin, stony, and infertile. Agriculture in the region is concentrated in river valleys and coastal lowlands such as the Minas Basin and Annapolis Valley of Nova Scotia, the Saint

Figure 1.2 *Panmure Island Provincial Park, Prince Edward Island.* [*Bill Brooks/Masterfile*]

John Valley of New Brunswick, and Prince Edward Island. Unlike the hard rock that underlies the hills, P.E.I. bedrock is sedimentary rock, and much of the Island is covered by a deep reddish soil that is especially well suited to growing potatoes and other root crops.

The other pervasive presence is the sea. Much early settlement in all four provinces happened along the coast. But even for those who live inland, the sea is never far away or long out of mind.

Figure 1.3 *Cape Breton Highland, Nova Scotia.* [*Greg Stott/Masterfile*]

This is the geographical reality that both the aboriginal peoples of the region and the European immigrants of the 17th, 18th, and 19th centuries encountered. Most of these immigrants came from the British Isles, and most found their livelihood in fishing, farming, forestry, or some combination of these. Although Canada as a whole is ethnically diverse, the Atlantic region is less so. Over 60 per cent of the residents of P.E.I. and Nova Scotia are of British ancestry, as are more than 85 per cent of Newfoundlanders. All four provinces also have Acadian populations, descended from the early French settlers in the region. And all still have aboriginal populations. People of the Mi'kmaq First Nation live in all four provinces. The Maliseet live in New Brunswick, the Shanadithit in Newfoundland, and the Innu and Inuit in Labrador.

While there are distinct differences between the four provinces and the cultural groupings within each, there is everywhere a deep regard for community and friends. Music — much of it rooted in the Scots/Irish and Acadian musical traditions — has always been important in this part of the world. For many, music is part of everyday life. People make music as well as listen to it, and they carry their instruments along to the Saturday night party. Literature and art are also alive and well in Atlantic Canada. Much of the region's literature is grounded in aspects of local culture, and much of its art celebrates the region's natural, sometimes rugged, beauty.

The Economic Reality

The good news is that most people who live in Atlantic Canada love it. The bad news is that it's tough to make a living there. In 1992, for instance, the per capita gross domestic product for Nova Scotia was just 80% of the national figure. In the other three Atlantic provinces it was even lower: New Brunswick 77%, P.E.I. 69%, and Newfoundland 66%. The seasonal nature of key economic activities — farming, fishing, and tourism — is one reason. The costs and delays associated with being a long way from major markets is another.

The Bridge and the Economy

Abiding in the hearts of most Canadians is a willingness to help those neighbors of ours who are having a tougher time than we are. It is manifested in our philosophy of universal health care and in the concept of *equalization payments*, where the Federal government redistributes tax revenues from the wealthier provinces to less wealthy ones. Associated with the equalization payments made to Atlantic Canada are various federal government activities aimed at promoting the economic development of the region. Major initiatives in this area have included the 1969 *Comprehensive Development Plan for Prince Edward Island*, and the formation in 1987 of ACOA, the *Atlantic Canada Opportunities Agency*.

In line with this economic development philosophy, Public Works Canada wanted Atlantic Canada to receive major economic benefits from construction of the Confederation Bridge. The agreement with the Bridge developer, Strait Crossing Development Inc., mandated that 70 per cent of total procurement for the project and 96 per cent of the labour would come from Atlantic Canada. Between 1993 and bridge opening on June 1, 1997, the project put more than $700 million directly into the economy of Atlantic Canada. Efforts were also made to see that certain unavoidable negative economic impacts would be dealt with fairly and equi-

Figure 1.4 *Bauline, Newfoundland.* |*Bill Brooks/Masterfile*|

Figure 1.5 *North Head, Grand Manan Island, New Brunswick.* [*Bill Brooks/Masterfile*]

tably. A $10 million fund was set up to compensate those in the fishing industry whose livelihood was temporarily affected by bridge construction activities. Steps were also taken to ensure that the approximately 420 displaced ferry workers received fair and equitable treatment. These steps included

- first choice for the 40 to 50 operation and maintenance positions on the bridge,
- a negotiated severance package, and
- retraining and relocation assistance.

In addition, $20 million was allocated to pursue development opportunities in the two communities most affected by the new bridge — Borden-Carleton, P.E.I., and Cape Tormentine, New Brunswick. The Federal government also provided $20.4 million each to New Brunswick and P.E.I. to defray highway system costs connected with the bridge.

Because of its distance from Central Canada, Eastern U.S. cities, and Europe, transportation always has been — and still is — a big issue in Atlantic Canada. Back in the second half of the 19th century when railroads were king and Canada was just forming, the Intercolonial railway was built to connect Halifax, Saint John, Moncton, and several other Maritime cities with Central Canada. Freight rates were subsidized, and this gave Maritime goods a chance to compete in Central Canadian markets. At the same time, the Prince Edward Island transportation situation was quite bleak. P.E.I. would build its own (heavily-indebted) railroad during this period, but there was no way of connecting it to the mainland rail system. In fact, P.E.I. was almost completely isolated for much of each winter. Even transporting people and mail between the Island and the mainland was a severe trial. The next chapter tells the story of how Prince Edward Islanders have dealt with this problem during the past two centuries.

CHAPTER TWO

BOATS AND ICE

In summer, transportation between Prince Edward Island and the mainland has always been manageable. Given a fine summer day, a good-sized canoe, and a few strong paddlers, people made it across the Northumberland Strait. When sail-powered boats came along it became easier still. The larger sailing vessels were able to carry many people and substantial amounts of cargo. The problem was getting across in the winter.

Ice Boats

There were Mi'kmaqs who occasionally crossed the Strait in mid-winter. But it was the Colonial government's desire to have year-round mail service that led to the first such trips in more recent times. In February, 1775, Governor Walter Patterson persuaded some Islanders to carry the mail across the ice and open water by canoe from Wood Islands at the southeastern end of P.E.I. (then St. John's Island) to Pictou, Nova Scotia. The trip succeeded, and two years later (in the absence of Governor Patterson who had been taken prisoner by American privateers) Acting Governor Phillips Callbeck tried again. He noted that "if this second attempt succeeds, it

Figure 2.1 *For propulsion across the ice, sails sometimes helped. But over the many rough places, the boat had to be hauled.* |PAPEI|

Figure 2.2 *Male passengers were required to help haul the boat across the ice, but still had to pay $2 per crossing. Women, invalids, and elderly passengers stayed in the boat but paid twice as much.* [PAPEI]

Figure 2.3 *Put in service in 1876, the 144 foot, 700 hp* Northern Light *was sometimes ice-bound for days, and proved better at breaking ice when it went backwards than when it went forwards.* [PAPEI]

will be a means of removing an objection which many people have made against living here, they being so long shut up from any intercourse with the rest of the world."

Trips via the 23-mile route from Wood Islands, to Pictou Island, to Pictou continued at least sporadically for the next fifty years, but few details are available. The Mi'kmaqs had preferred the much shorter 9-mile passage from Cape Traverse on the Island to Cape Tormentine on the mainland — close to where the Confederation Bridge now crosses the Strait. Lt. Governor John Ready wanted to explore this route, and on December 19, 1827 Neil Campbell and Donald McInnis made a successful crossing. In 1828 some mail was sent by this "Capes Route," and in 1829 the Island government arranged for all mail and passengers to go that way.

During a long, exceptionally bitter winter, the Strait's ice cover might grow until it was nearly continuous, and there are occasional reports of someone walking all the way across. Almost always, however, there was a mix of bord (landfast) ice near the shore, open water, and floating ice floes. This meant that a boat could be rowed or sailed for part of the journey, but for the remainder it had to be hauled across the ice by crew and passengers. A canoe was far from ideal in these circumstances, and over the years an optimized "ice boat" design evolved. We don't know much about the experiments tried along the way, but we do know about the design that proved most effective. The classic ice-boats of the mid-1800s were 17 or 18 feet long, 4 or 5 feet wide, and about two feet deep. Their hulls consisted of bent cedar planks attached to an oak frame, and were generally clad with heavy tin. The tin protected the wood from abrasion when the boat was hauled over ice hummocks and ridges, and it helped reduce friction. Iron-clad wooden runners were attached to the boat on both sides of the keel, and the boat rode on these while being pulled across the flatter stretches of ice.

For propulsion in open water, ice boats were equipped with oars and a small sail. For propulsion across the ice, the sail sometimes helped. But over the many rough places, the boat had to be hauled. Each ice boat was equipped with long leather straps that went over the shoulders of the men doing the hauling; they transferred the pulling force from man to boat. Some boats were also equipped with harnesses or waist-belts for the purpose of saving the hauler if the ice broke under his feet — something that happened fairly frequently. Legislation passed in 1834 required that ice boats be at least 16 feet long and carry no more than 4 passengers; it also limited baggage to 20 pounds maximum per passenger.

During the 1830s the typical iceboat would have a captain and a crew of three, and would make one trip per week over and back. The 9-mile passage across was normally made in one day, but if a bad storm arose, or especially difficult ice conditions were encountered, the situation for passengers and crew became life threatening. Many of those who were forced to spend a night on the ice in a raging storm lost fingers, toes, hands, or feet to frostbite, and some lost their lives to hypothermia.

One of the worst disasters occurred in 1855. The morning of Saturday, March 10 was sunny, and the winds at Cape Tormentine were light. Captain McRea, his crew of three, the mail, and three passengers left at 8 a.m. for Cape Traverse, P.E.I. One of the passengers was Joseph Weir, an older gentleman from Bangor, Maine who had his pet spaniel with him. The other passengers were Harry Hazard and Richard Johnston, medical students on their way home to Charlottetown from the United States. Later

Figure 2.4 *Shown here stuck in the ice, the 207 foot, 1900 hp, steel-hulled* Stanley *replaced the* Northern Light *in 1888. It was a better ice-breaker, but still no match for* Northumberland Strait *ice at its worst.* [PAPEI]

Figure 2.5 *A stronger, heavier, ship, the 2900 hp* Minto *joined the* Stanley *in 1899, but it, too, sometimes got stuck in the ice.* [PAPEI]

in the morning there was a light snow shower, but the sun was soon out again. Then, in the afternoon, the wind picked up and shifted to the northeast.

By 5 p.m. the party was within two miles of their destination when they were stopped by a combination of "lolly" ice — ice that you couldn't stand on or easily row through — and a violent snow and sleet storm. They pulled their boat up on an ice floe. The blizzard continued through the night, with the boat providing little protection from the wind. By Sunday morning the passengers were frostbitten. Their ice floe had traveled a considerable distance during the night. The Island was evidently some distance away now, and the group decided that it would be best to head back toward the mainland. They struggled on through Sunday, and spent another night on the ice. By noon Monday the crew was exhausted and so hungry that they decided (reluctantly we are told) to kill Mr. Weir's spaniel, drink its blood, and eat its raw flesh. This revived the crew, and after discarding the luggage but keeping the mail, they continued on through the rest of Monday.

Meanwhile young Mr. Hazard became delirious and refused to leave the boat. They covered him with blankets, but that evening he died of hypothermia. The others kept on, pulling the boat with its cargo of corpse and mail. About midnight the party reached land at Fox Harbour, Nova Scotia — some twenty-five miles east of where they had started on Saturday morning. Leaving the boat and Hazard's frozen body on the shore, the others headed through the deep snow. At 4 a.m. McRea, Johnston, and one crew member reached a farmhouse; the rest had fallen behind. Two occupants of the house headed out to look for the missing men and eventually found them lying in the snow — exhausted but still alive. All but one of the survivors had badly frostbitten feet, and Weir would lose his toes — but they had made it through the ordeal.

As the 19th century progressed, these ice boat disasters were taken to heart by the public and governments, and safety slowly improved. One improvement was to have two or more boats travel together, and this became economically practical as the volume of mail increased and more people made winter crossings. In 1885 (following another disaster) the Dominion government decreed that each boat must carry survival basics such as 4 oars, a compass, food, axes, and fire-making materials. In addition, each boat now had to have a crew of 6 (including the captain), and crossings were to be made in a fleet of no less than three boats.

The fact that male passengers were required to help haul the boat across the ice didn't mean that they had free passage. Their fare was $2 per crossing. Women, elderly passengers, and invalids stayed in the boat and were charged $4. A gentleman would be entitled to stay in the boat if he paid $5 — but when the going got rough he, too, would be expected to get out and pull. These were not trivial amounts of money in those days. As late as 1913, $6.34 could buy a week's food for a P.E.I. family, and $1.75 was a week's rent.

Figure 2.6 *The* Prince Edward Island, *was the first railcar ferry. It began service between Borden and Cape Tormentine in 1917, and in the first year of operation made 506 round trips. The improvement in reliability was so dramatic that the ice boat service was permanently discontinued and the* Stanley *was taken out of service.* [PAPEI]

Crew members were paid $2 per round trip, but they had to spend a night on the other side, and food and lodging there cost them $1. The captain's wage was $5 for each round trip. Typical crew clothing included heavy winter underwear, pants or overalls, a sweater, a jacket, and rubber boots. The boots came above the knees, and had screws driven into the soles to give good traction on the ice.

Efficient Steam Service

Throughout the late 1860s and early 1870s the Island government and the newly formed Dominion of Canada negotiated terms for P.E.I.'s entry into Canada. One of the main issues was transportation between the Island and the mainland, and on May 20, 1873 Parliament petitioned the Queen to admit Prince Edward Island into the Dominion of Canada under conditions which included:

> "the Dominion Government shall assume and defray all the charges for the following services, viz.: —
>
> Efficient steam service for the conveyance of mails and passengers to be established and maintained between the Island and the mainland of the Dominion, winter and summer, thus placing the Island in continuous communication with the Intercolonial Railway and the railway system of the Dominion; . . ."

This became part of the Canadian constitution.

The words were wonderful, and created great expectations. But the reality which unfolded over the next forty-four years was something else. The Dominion Government first tried to meet the winter part of its obligation by having the old wooden-hulled *Albert* do winter runs between Georgetown and Pictou. It proved totally unable to cope with Strait ice. In December of 1876 the government put the 144 foot, 700 hp *Northern Light* on this run — also with disappointing results. It was

Figure 2.7 *The original* Abegweit *or* Abby *captured the hearts of Prince Edward Islanders. Put into service in 1947, it was at that time the largest, most powerful icebreaking ferry in the world. In 1982 it was replaced by a new* Abegweit, *and now resides at Chicago's Lake Michigan waterfront as the clubhouse of the Columbia Yacht Club.* [Marine Atlantic]

frequently damaged by the ice, was sometimes ice-bound for days, and proved better at breaking ice when it went backwards than when it went forwards. In 1881 it was ice-bound for three weeks, and the Island government complained that the service being provided was neither efficient nor continuous.

The Dominion government did nothing, and as the 1880s wore on, Island outrage increased. Ice boat service was improved and made safer during this period, and in 1887 P.E.I. was granted an annual subsidy of $20,000 to help with its transportation problem. Still, it wasn't till 1888 that the highly unsatisfactory *Northern Light* was replaced by a new ship. Its replacement was the *Stanley*, named after Lord Stanley — the Stanley Cup's namesake. The *Stanley* was a 207 foot, 914 ton, 1900 hp, steel-hulled, ice-breaking ship, and hopes were high that this was finally the answer to the Island's winter transportation problem.

Figure 2.8 *The second* Abegweit *was the largest and most powerful of the* Northumberland Strait *ferries. Put in service in 1982, this ferry could carry 974 passengers and 250 automobiles.* [Marine Atlantic]

In the winter of 1888-89 the *Stanley* made 79 round trips compared to 21 round trips made by the *Northern Light* the winter before. Islanders were delighted, but their delight was short-lived. The following winter was a harsh one, and on 43 days the *Stanley* was unable to cross the Strait. It was a much better icebreaker than the *Northern Light*, but was still no match for Northumberland Strait ice at its worst. The Dominion government tried twice more to get it right — first with the *Minto* in 1899 and then with the *Earl Grey* in 1909. The Earl Grey was an improvement, but as part of Canada's contribution to the war effort, both ships were sold to Russia in 1914 . The *Stanley* was put back in service, and ice boats continued to be used until 1917 to get the mail across when the *Stanley* couldn't.

Throughout this period the Island government complained that the Dominion government was not living up to the terms of union. Acknowledging that this was in part true, but at the same time not having a real solution, the Dominion government raised P.E.I.'s 1887 annual subsidy of $20,000 by an additional $30,000/year in 1901, and in 1912 by $100,000/year more. While this was appreciated, it was not an adequate substitute for transportation that really worked. Not until 1912 when the Dominion government under Robert Borden decided to build a railcar ferry to run between Carleton, P.E.I. (later renamed Borden) and

Cape Tormentine, N.B. was satisfactory winter transportation in sight.

The first railcar ferry, the *Prince Edward Island*, was 300 feet long, rated at 7000 hp, and could carry 9 railcars and 750 passengers. In addition to having two propellers aft, it had a bow propeller to help break ice. During 1915-16, before the Borden and Cape Tormentine terminals were completed, it sailed between Charlottetown and Pictou. In 1917 the *Prince Edward Island* began service between Borden and Cape Tormentine, and in its first year of operation made 506 round trips. The improvement in reliability was so dramatic that the ice boat service was permanently discontinued, and the *Stanley* was taken out of service.

With this change, P.E.I.'s ferry service had come closer to that elusive goal of "Efficient Steam Service . . . Winter and Summer" and "continuous communication with the . . . railway system of the Dominion." In 1938, in response to the growing use of automobiles, the *Prince Edward Island* was equipped with an auto deck. And it continued to be used as a ferry until 1969 — an amazing 54 years.

A privately owned warm-weather ferry service between Wood Islands, P.E.I. and Pictou, Nova Scotia began operation in 1941 and still continues. Over the years, new ferries were added to the government operated system between Borden and Cape Tormentine, and older ones were retired. The accompanying table notes these events, and gives some of each ferry's characteristics. We can't close this chapter, however, without saying a few words about one special ferry that captured the hearts of Prince Edward Islanders and many Island visitors: the original *Abegweit*, the *Abby*. Put into service in 1947, it had roughly twice the tonnage and twice the horsepower of the *Prince Edward Island*. It was the heaviest vessel that had ever been built in Canada up to that time, and was the largest, most powerful icebreaking ferry in the world. Its passenger areas were elegantly finished with beautiful mahogany, walnut, and oak woodwork, brass hardware, decorative frosted glass, and inviting upholstery. In fact, its decor was more like that of an ocean liner than a ferry, and white-jacketed waiters served meals in its dining room. During its 35 years of service it crossed the Strait 123,207 times, sailing 1,145,585 miles in the process. In 1982 it was replaced by another ship bearing the same name. The original *Abby* now resides at Chicago's Lake

Figure 2.9 *With ironic timing, a windstorm on December 20, 1996 drove the Marine Atlantic ferry* John Hamilton Gray *onto a sandbar where it remained stuck for two days. The Confederation Bridge structure had been completed just a month before, but it would be another five months before the bridge was ready for traffic.* [George Read]

Michigan waterfront as the clubhouse of the Columbia Yacht Club. Gone from the Strait, but still loved and not forgotten.

The 1873 terms of union referred to "Efficient Steam Service," and from then until very recently there has been public pressure to improve the ferry service across the Strait. From time to time, however, people have proposed various fixed-link alternatives to crossing by boat: subways, tunnels, causeways, and bridges. The next Chapter tells that story.

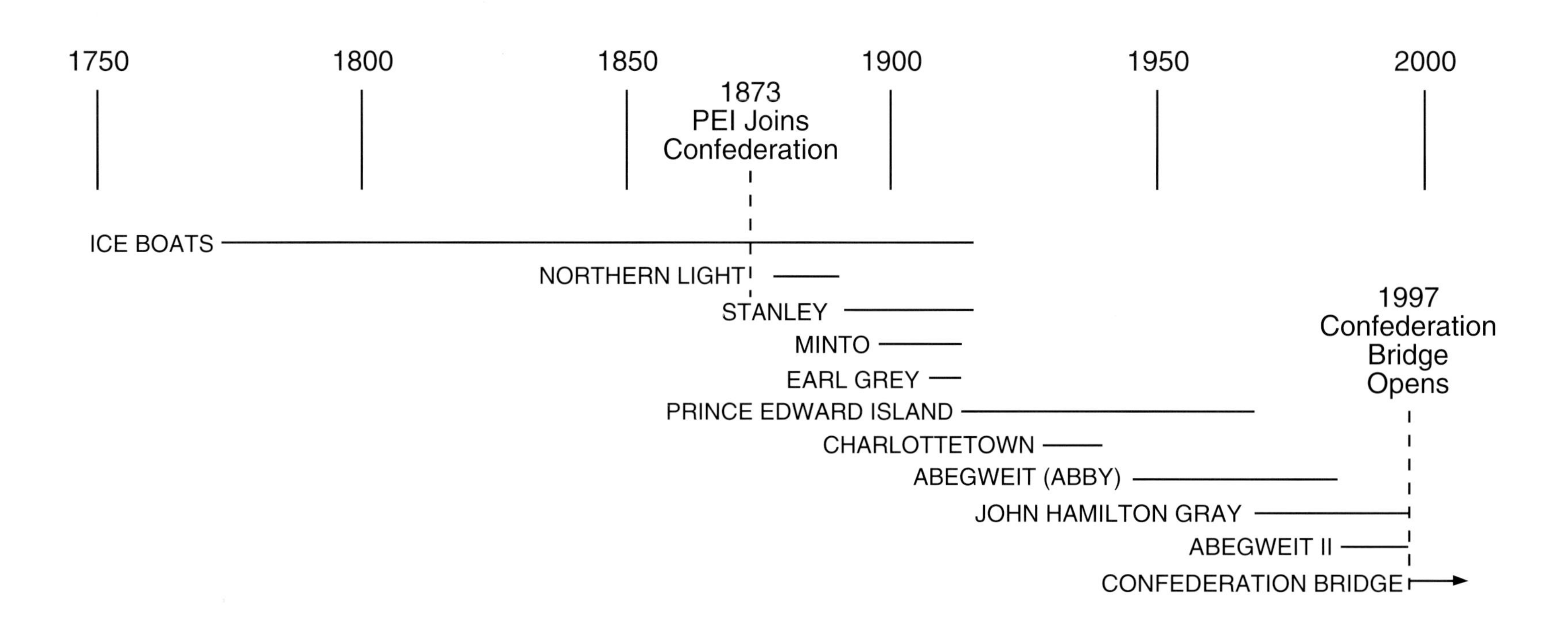

ICE CONDITIONS IN NORTHUMBERLAND STRAIT

Ice conditions in the Strait are notoriously variable. In the ice boat days people encountered Strait ice first hand, and their lives depended on understanding the differences between its many forms. Because of this, a complex vocabulary arose around ice conditions. A few of the most commonly mentioned terms were:

- Bord Ice — continuous ice that is attached to the shore and extends out for a distance ranging from a few tens of yards to more than a mile. Today, this type of ice is often called landfast ice.
- Hard Ice — floating ice pans or floes with an average thickness ranging from 10 or 12 inches to 3 feet or more. Many floes have hummocks and ridges, sometimes 20 feet high or higher.
- Raft Ice — a floating ice formation resulting from the piling up of cakes of ice in layers.
- Lolly Ice — slush mixed with sizable chunks of hard ice. Lolly ice was impossible to walk on, and almost impossible to row through. Three-pronged boat hooks were sometimes used to pull the boat past the chunks.
- Soft Ice — ice that was mostly slush, mixed with small hard chunks. Ice boat crews could row through it.

Then and now, the first ice to form in the Strait is normally bord or landfast ice along the shores of estuaries and harbours — and it usually starts to form in December. As winter progresses, landfast ice develops elsewhere along the shore of the Strait, and it extends further out from shore. At some point, usually in January, ice begins to form on the surface of open water away from land. The initial thin layers are broken up by tide, wave, and wind action, and they slide over one another and refreeze. Gradually, the ice becomes thicker and consolidates into ice floes that are moved along the Strait by currents and the wind. Additional floating ice enters the western end of the Strait from the Gulf of St. Lawrence. Mid-winter floes range in size from about 200 feet in diameter to over a mile, with most falling in the 500 to 1500 foot range. Wind and tide move floes in various directions, but there is a general long-term movement from west to east.

When storms and tides drive floes against, over, and under each other — or into landfast ice — hummocks and pressure ridges many feet high can form. This dance of ice proceeds through the winter. Floes break apart, consolidate, and grow in height and depth. Yes, there are downward-going ridges too, often much larger than the upward-going ones. These can scrape or "scour" the bottom of the Strait, and scour trenches in water as deep as 45 feet have been observed.

Eventually, the lengthening days of late winter cause the ice-growing process to reverse itself. Floes get smaller, the water gets warmer, and eventually "ice out" day arrives — the day when the last of the winter ice has disappeared from the Strait. Given a normal winter, ice out occurs sometime between April 2 and May 31, with the average being April 30. If the winter has been especially warm or cold, this cluster of dates will shift a few days earlier or later, but regardless of the severity of the winter there is always a wide spread between the earliest possible ice out date and the latest one.

THE BORDEN / CAPE TORMENTINE FERRIES

ICE-BREAKING AND SUMMER-ONLY BOATS IN SERVICE FROM 1917 TO 1997

Name of Ferry	Start/End of this Service	Ice Breaker?	Length (Feet)	Total Power (Horsepower)	Passenger Capacity	Rail-Car Capacity	Automobile Capacity
Prince Edward Island	1917 - 1969	yes	300	7,000	750	9	36†
Scotia I*	1920 - 1953	no	272	3,000	—	16	55
Scotia II*	1920 - 1953	no	272	3,000	—	16	55
Charlottetown	1931 - 1941	yes	324	8,000	800	16	45
Abegweit (I) — Abby	1947 - 1982	yes	372	13,500	800	16	45
Confederation	1962 - 1975	no	283	10,200	400	—	60
John Hamilton Gray	1968 - 1997	yes	400	16,000	516	18	165
Lucy Maude Montgomery	1969 - 1973	no	262	6,000	400	—	90
Holiday Island	1971 - 1997	no	325	6,000	485	—	155
Vacationland	1971 - 1997	no	325	6,000	485	—	155
Abegweit (II)	1982 - 1997	yes	402	22,000	974	20	250

† Originally designed to carry only rail cars, an automobile deck was added in 1938.

* The Scotia I and Scotia II were just powered barges, with no passenger amenities. They were built to ferry railcars across the 1-mile-wide Strait of Canso. Still, they occasionally served as summer relief boats on the Borden / Cape Tormentine run when one of the regular ferries had to go into dry dock for servicing or there was a freight backlog.

CHAPTER THREE

EXPLORING ALTERNATIVES

In the view of Islanders, the Dominion government's 1873 promise of continuous steam service to the mainland was not being kept — and they were mightily upset. Too much of the time the so-called ice-breakers weren't breaking the ice, and many Islanders must have dreamed of a tunnel or bridge that would get them safely across the Strait any day of the year. Great tunnels were being built in Europe at the time, and great bridges in North America; couldn't one or the other be built between P.E.I. and New Brunswick?

Senator Howlan's Subway Proposal

The first Islander to persistently press for an alternative of this kind was Senator George W. Howlan of Alberton, P.E.I. Senator Howlan had helped draft the 1873 terms of union, and in 1885 he began actively promoting the construction of what he called a *subway* — a long hollow structure that would lie on the seabed at the bottom of the Strait. As described in a speech he made to the Senate in April of that year, the subway would consist of five miles of interconnected iron tubes. Each tube would be 300 feet long and 15 feet in diameter. The interior of the tubes would be lined with a 2-1/2 foot thick layer of concrete, and the concrete at the bottom of the tube would support a railway roadbed. (Five miles of subway would be enough, he figured, because long earth-filled wharves would be built out from each shore.)

Senator Howlan noted that the Thames River in London already had two such subways crossing it — though these were much shorter than the one he was proposing. One of the Thames tunnels carried narrow gauge trains, the other standard gauge. Senator Howlan quoted Vernon Smith, a Canadian engineer, who declared that the project was feasible and estimated its cost at $2.1 million.

Howlan had looked into the feasibility of a bridge, but ruled it out. He concluded that any bridge that could be built across the Strait would be so close to the water that it would seriously interfere with navigation. Also, using the bridge-building techniques of the day, part of each summer's progress would be torn away by the following winter's ice.

The Senator also looked into the feasibility of a tunnel carved through the bedrock under the Strait, but concluded that this was not the way to go either. Engineer Vernon Smith cited the problem of ventilating so long a tunnel and the likelihood of

Figure 3.1 *Senator George W. Howlan of Alberton, P.E.I. had helped draft P.E.I.'s 1873 terms of union. In 1885 he began to actively promote the construction of what he called a* subway *— a long hollow structure that would lie on the seabed at the bottom of the Strait and allow railroad trains to pass between Prince Edward Island and New Brunswick.* [PAPEI]

water seepage. One speaker in the Senate noted that the tunnel being built under the Severn estuary to connect England and Wales had been going well until the workmen encountered a spring. "They were driven out so completely that the men had some difficulty in escaping with their lives, and the horses were actually drowned. The tapping of that spring retarded the work for several years."

The difference between tunnel and subway appears to have been a subtle one in the public mind, and the term tunnel was frequently applied to both types of construction. Even Senator Howlan apparently bowed to that reality when that same year he formed The Northumberland Straits Tunnel Railway Co. The name on the company's letterhead was Tunnel, but it pictured his sectionalized subway.

In March of 1886 a delegation from P.E.I. went to London to present a memorandum to the Colonial Secretary complaining that the Dominion Government was not meeting its obligation under the terms of confederation, and that perhaps a railway tunnel was the solution. The British government refused to get involved.

Back at home, P.E.I.'s Premier said "there is every reason to believe that Sir John Macdonald's government will carry out the terms of Union with this Province, and that we shall have either a subway or a tunnel across the Straits." Howlan, too, appealed to the Prime Minister. That summer Canadian government engineers examined the bottom of the Strait between Cape Traverse and Cape Tormentine, but the government took no further action.

Senator Howlan continued to press for a tunnel until his death in 1901. Afterward, Howlan's friend Father A.E. Burke championed the cause. Over the years mass meetings were held, delegations of Islanders went to Ottawa, and in the great Canadian way, governments studied the matter very carefully. No government ever committed money to build a tunnel. In 1912, Prime Minister Robert Borden announced that his government had decided to establish a railcar ferry service. The ferry *Prince Edward Island*, by linking the mainland and Island railway systems, met the central transportation need of the day. Tunnel talk all but ended.

The 1960s Causeway Project

In 1928 the federal Department of Railways took one more look at the cost of building a tunnel, and also the cost of a new approach — a *causeway*. This would be "a breakwater structure with a road on top and adequate facilities here and there for the passage of boats to and fro." There were doubts about whether a causeway could survive Strait ice conditions, and with the high cost of both projects — estimated to be $40 million for the tunnel and $46 million for the causeway — both were dropped.

In the early 1950s a causeway and short-span bridge were built across the 1-mile-wide Strait of Canso that separates Cape Breton Island from mainland Nova Scotia. The project was completed in 1955, and it stimulated a decade of causeway interest in nearby P.E.I. O.J. McCulloch had designed the Canso Causeway, and the P.E.I. government under Premier A.W. Matheson asked him to estimate the cost of a Northumberland Strait causeway. McCulloch's preliminary estimate was $50 million. He also noted that the Canso Causeway had required nearly 10 million tons of rock; a P.E.I. causeway would require at least 40 million tons.

In the years that followed, federal government departments studied the matter from many sides, and P.E.I. kept pushing for action. In 1962, with an election on the horizon, Prime Minister Diefenbaker announced that the project was technically and economically feasible, and that the federal government would supply the $105 million needed to build the causeway. Shortly after, a consortium of Canadian consulting engineers began reviewing past studies, conducting additional studies, and working on the conceptual design. They concluded that the best approach would be a combination bridge, causeway, and tunnel that would accommodate both road and rail traffic.

In the election campaign of 1963, both parties supported the project. Diefenbaker was defeated and Pearson was elected, but design and other pretendering work proceeded. Finally, on July 8, 1965, Prime Minister Pearson announced that the government would move forward on the construction of the combined causeway, tunnel, and bridge. The first contracts were awarded later that year, and on November 5th — a few days before a federal election — sod was turned on the P.E.I. highway approach.

Pearson was reelected, and work continued on the approach roads. Tendering for construction of the causeway itself came later. In 1967, bids for the causeway on the New Brunswick side were received but rejected. They came in much higher than expected. Concerned, the federal government decided to reassess the project. It commissioned studies on alternative crossing schemes, and on the costs and benefits of a fixed crossing.

By 1968 the federal government had spent $15 million on the project. Earth-moving work for the New Brunswick approach road had been completed, and the Albany Interchange on the P.E.I. side would soon be finished — but all was not well with the project. In the spring of 1968 the causeway project's yearly expenditure was cut to $5 million, and there were rumors that the project might be cancelled. Pierre Trudeau became Prime Minister that year, and there were new government initiatives. One of these involved discussions between the new federal Department of Regional Economic Expansion and the P.E.I. government about economic development on the Island.

On March 5, 1969 the project *was* cancelled. On that date the federal government announced that instead of funding the causeway it would contribute $225 million over the next 15 years toward a *Comprehensive Development Plan* for the Island. Prime Minister Trudeau apologized for not being able to fund both the Development Plan and a causeway. He said that his government decided "to support the Development Plan as being the likeliest method of offering appreciable and lasting benefit to the economy of Prince Edward Island in the foreseeable future." The causeway project was history. Considering what we now know about the environmental problems that causeways produce, it's no doubt a good thing.

Renewed Interest in the Mid-1980s

During the 1970s and early 1980s there were improvements to the ferry service, and little talk about a fixed crossing. Two summer-only ferries (the

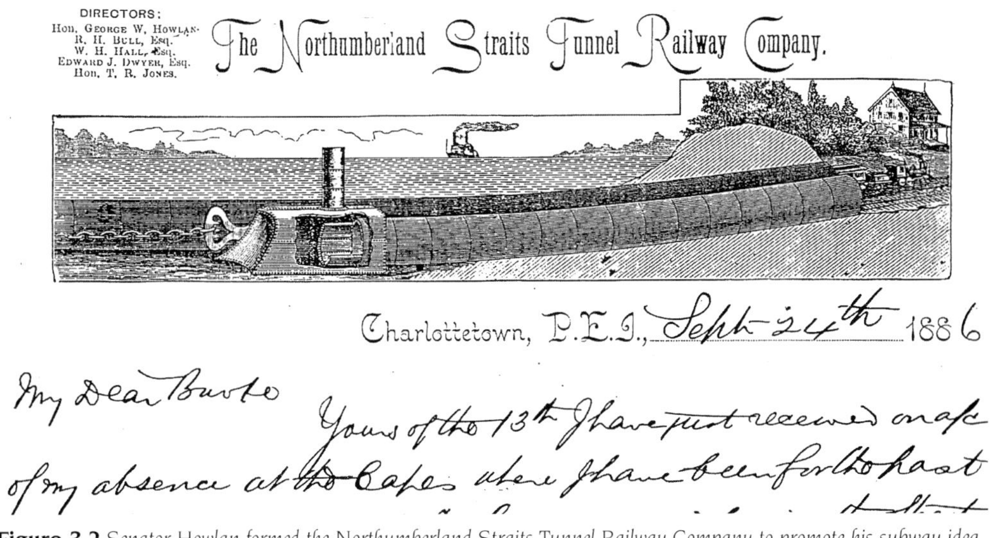

DIRECTORS:
Hon. George W. Howlan.
R. H. Bull, Esq.
W. H. Hall, Esq.
Edward J. Dwyer, Esq.
Hon. T. R. Jones.

The Northumberland Straits Tunnel Railway Company.

Charlottetown, P.E.I., Sept 24th 1886

My Dear Burke

Yours of the 13th I have just received on a/c of my absence at the Capes where I have been for the past

Figure 3.2 *Senator Howlan formed the Northumberland Straits Tunnel Railway Company to promote his subway idea. In an 1885 Senate speech he described the subway as consisting of five miles of interconnected iron tubes. Five miles of subway would be enough, he figured, because long earth-filled wharves would be built out from each shore.* [Hubert Jacquin]

Figure 3.3 *Early button promoting the tunnel.* [PEIM&HF]

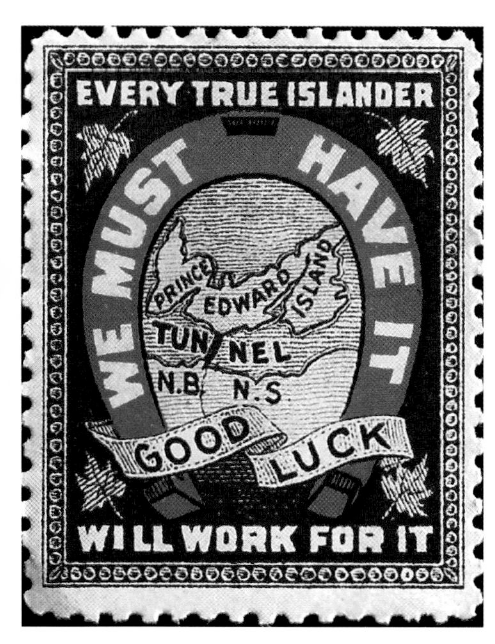

Figure 3.4 *Early tunnel-promoting stamp.* [PAPEI]

Vacationland and *Holiday Island*) helped relieve the summer tourist crunch, and the original *Abegweit* was replaced by a much larger all-season ferry bearing the same name. At the same time, the ferry system became increasingly expensive to run. Ferry tolls rose faster than inflation, and the federal government kept increasing the amount of its subsidy.

In 1982, Public Works Canada (PWC) reviewed a proposed bridge design and estimated its cost at $640 million in 1982 dollars. Then, in 1985-86, PWC received three unsolicited proposals from the private sector. Omni Systems proposed a tunnel. Northumberland Bridge Builders proposed a bridge. Nova Construction first proposed a causeway/tunnel/bridge; it then became Abegweit Crossing Ltd. and proposed a bridge. A unique aspect of these proposals was that the projects would be *privately financed*. The federal government was, by now, putting out tens of millions of dollars per year to subsidize the Borden – Cape Tormentine ferry service. The basic idea was to have private-sector loans finance construction, then use ferry subsidy money to pay back the loans.

The government took notice. A major infrastructure project that wouldn't increase government debt and would eventually end the ferry subsidy was worth looking into. In December of 1986 the government gave Public Works Canada the mandate to examine the technical, financial, environmental, social, and economic dimensions of the proposed fixed-link concepts. PWC was also asked to assess private sector interest in financing, building, operating, and eventually transferring the facility to the federal government.

To answer a variety of questions, PWC commissioned ten studies. The studies probed such topics as the physical oceanography (including ice conditions), fishery resources and impacts, socio-economic impacts, vessel traffic and bridge safety, and tunnel feasibility. The reports on these studies were delivered during the spring of 1987, and on May 12, PWC called for "Expressions of Interest."

As conceived at that time by Public Works, the project would be divided into six stages:

I. Call for Expression of Interest (submit qualifications)
II. Invitation for Proposals (submit facility concept solution, environmental and socioeconomic assessment and plan, management approach, financial plan, developer's team capability and experience)
III. Selection of final Proposal (submit bid and final financial plan)
IV. Design
V. Construction
VI. Operation and Maintenance

The purpose of the Stage I Call for Expressions of Interest was to prequalify a short list of proponents to participate in the Stage II proposal call process. In their responses, these organizations had to show that they had sufficient financial, management, engineering, construction, operation, and maintenance capability to successfully execute the project.

The "Call" document defined the project and objectives in broad terms. "The project is a fixed crossing of the Northumberland Strait between Jourimain Island, New Brunswick and Borden Point, Prince Edward Island to be financed, designed, constructed, operated and maintained by the private sector under a lease purchase agreement for a specific period of time (tentatively 35 years)." It would then be turned over to the Federal Government. The document specified a 13 km main structure, 14 km of approach roads, and certain other parameters such as road width, clearances, etc. The fixed link was to have a useful life of 100 years, and there were two acceptable forms that it could take:

1. "A high level bridge structure from shore to shore across the Strait with provision for a navigation channel," or
2. "A tunnel for vehicular traffic."

Causeways, either all the way across, or part way, would not be allowed — nor would any sort of "immersed tube tunnel." (Alas, Senator Howlan loses again.)

The stated project objectives were:

1. **Private sector participation and involvement** — "the project will be financed, designed, constructed, operated, and maintained by the private sector (developer) . . ."
2. **Maximization of economic benefits to Canada and the Atlantic Region** — "It is the Government of Canada's intention that this project will be a significant contribution to the development of the Atlantic Region and . . . proposals will be evaluated on the basis of their contribution to the local economies."
3. **Value for money** — "an acceptable facility at the lowest possible cost to Canadians."

Twelve organizations responded by the deadline of June 5, 1987. Public Works Canada opened a P.E.I. information office that month, and Islanders began discussing, in earnest, the pros and cons of such a project.

PWC wanted to find out if the project had any fatal environmental flaws, and so commissioned an Initial Environmental Evaluation (IEE) on its loosely-specified "generic" tunnel and bridge. This generic IEE or GIEE would look at the effect of both kinds of structures on the marine, terrestrial, and socio-economic environment before asking the shortlisted firms to submit their Stage II proposals. It wouldn't be the final environmental word. If the GIEE did not rule the project out, the plan was to call for Stage II proposals, receive and evaluate them, and then have the successful bidder subject its specific design to a second IEE.

The consultants worked on the IEE through the summer, and in November issued a draft report. Public Works held information sessions in P.E.I., New Brunswick, and Nova Scotia to explain the results of the studies. Then, on November 16, Public Works Minister Stewart McInnis announced that "these studies have indicated that there are no environmental impacts that cannot be mitigated by proper design and construction procedures." He gave the green light for a proposal call early in 1988 if P.E.I. supported the project. The government also announced the names of seven firms that passed initial screening and would be eligible to submit proposals.

The P.E.I. Plebiscite

On P.E.I., this decision to proceed brought the simmering debate over a fixed crossing to a fast boil. Many Islanders applauded the decision; many others were aghast. Within three weeks of the announcement, three highly significant events occurred.

Premier Joseph Ghiz announced that Islanders would be able to express their views on the fixed-crossing issue in a plebiscite to be held January 18. Plebiscites were something of an Island tradition. Four had been held earlier this century, all of them about the use and control of alcohol. The most recent, in 1948, abolished prohibition in the province. In the January plebiscite, the question would be: "Are you in favor of a fixed link crossing between Prince Edward Island and New Brunswick." The choices on the ballot would be "Yes" or "No."

To ensure open debate before the plebiscite, the Premier requested the University of Prince Edward Island to sponsor a series of public meetings. This it agreed to do, and through its Institute of Island Studies planned a series of 12 meetings in 11 Island communities to "provide a forum for discussion" and to allow Islanders to "voice their opinions and concerns." At each session there would be a neutral moderator, speakers *for* and *against*, and questions and comments from the audience.

The Tunnel Route
169
Cape Traverse *
Cape Tormentine *

Figure 3.5 *Ticket to promote tunnel.* [PWGSC]

Figure 3.6 *Postcard promoting Howlan's tunnel.* [PEIM&HF *and* PAPEI]

Early advocates of a tunnel produced a variety of promotional materials to help create and maintain interest in the project. The postcard was particularly clever. If you looked at the back of the card while holding it up to the light, you could see the train crossing the Strait.

Figure 3.7 *Reverse side of the post card.* [PEIM&HF *and* PAPEI]

The third significant event was the formation of Friends of the Island, a group whose stated purpose was "To provide a rallying point for concerned Islanders who believe that to proceed with a fixed-link at this time would be ill considered . . ." and "To promote effective, coordinated participation in the December and January meetings by those Islanders opposed to the present proposal for a fixed link."

The public meetings were held, and Friends of the Island participated. The largest of these gatherings was held four days before the plebiscite. It was attended by well over a thousand Islanders and moderated by CBC's Peter Gzowski. P.E.I. historian David Weale spoke against the fixed link. He expressed his feeling that the issue was being pushed too fast, and that Islanders weren't being given sufficient time for an informed assessment. He also focused on the "Islandness" issue. He said, "We are Islanders . . . living in a place set apart by time and nature. And though it is denied by some, it is our Islandness which is at the centre of this debate."

In supporting the fixed link, lawyer David Hooley focused on the issue of closure. He argued that voting yes did not necessarily mean that the project would go ahead, but that voting no would surely kill it. The Friends of the Island had been saying "If you don't know, vote no." The way Hooley and the pro-link Islanders for a Better Tomorrow saw it, "If you vote no, you'll never know."

Support for a fixed crossing came from people who saw potential economic benefits, or liked the increased convenience or simply hated the ferry. Most were drawn to one or more of the specific benefits: Building and operating a fixed crossing would cost less than operating a ferry system. There would be major time savings for users. Users would have improved access to mainland markets. A fixed crossing would provide a reliable, low-cost power and communication utility corridor. And the amount of fuel consumed by vehicles driving across the Strait would be much less than that consumed by a ferry carrying those vehicles. (The ferries consumed about 20 million litres of diesel fuel per year and carried close to a million vehicles across the Strait. That's 20 litres per vehicle for a 13 km trip.)

Opposition took many forms, but it seemed to fall into two general categories:

1. A gut-felt dislike of the idea of a fixed-crossing, a philosophical opposition to it.
2. Fears and concerns about the consequences of building a fixed link — environmental concerns, safety concerns, socio-economic concerns.

The first kind of objection was clearly expressed sometime later by Donald Stewart, one of the founders of Friends of the Island: "Many people are simply opposed to anything that would diminish the status of Prince Edward Island as an Island. For them it is a matter of the spirit, and no compromise or mitigation is possible. The construction of a fixed link would represent a psychic violation."

This first kind of objection was often mixed with elements of the second. A decade earlier, when Alex Campbell was Premier, there was a saying doing the rounds: "We're so far behind we're ahead." This was, in many ways, true. We Islanders lived in a place of natural beauty, rural charm, neighborliness, and relatively little crime. To what extent would a fixed link change that?

Some fears and concerns were quite specific. There were concerns about the economics of the project — both its economic feasibility and the economic impacts it would cause. Many, for instance, saw the loss of several hundred well-paid Marine Atlantic jobs as a blow to both the ferry workers and the P.E.I. economy. There were fears that wind and ice might take down the bridge. There were fears that the bridge would damage one or more of the Strait fisheries.

There were fears that the project might suffer the fate of the 1965 causeway and never be completed. And there was the fear that easier access to the Island would bring uncontrolled development and increased absentee ownership of Island land.

Plebiscite day came. Voter turnout was a respectable 65%, but less than the usual 85 to 90% typical of general elections on P.E.I. By the end of the evening the results were in — YES, 59.4%; NO, 40.6%. In a public statement that evening, Premier Ghiz said, "I think a majority of Islanders have expressed a desire to have a fixed link connecting P.E.I. and the mainland. . . . It is a clear mandate to negotiate with the federal government while respecting the concerns of the many Islanders who voted against it."

The Premier had already addressed several of the most important concerns. Back in November he had written to Stewart McInnis outlining what came to be known as "Joe's 10 Commandments." In his letter he said that his government would be willing to support the project if certain conditions were met:

1. financial support for connecting highways,
2. continuation of the Wood Islands ferry at the eastern end of the Island,
3. an agreement concerning the land under the Strait which the fixed crossing would occupy,
4. tolls (which had to be "fair and reasonable"),
5. compensation for displaced ferry workers,
6. economic development in Borden,
7. economic benefits to the Atlantic region,
8. the resolution of "all significant environmental issues,"
9. compensation for adversely affected fishermen during construction, and
10. a utility corridor within the tunnel or bridge through which cables could be run.

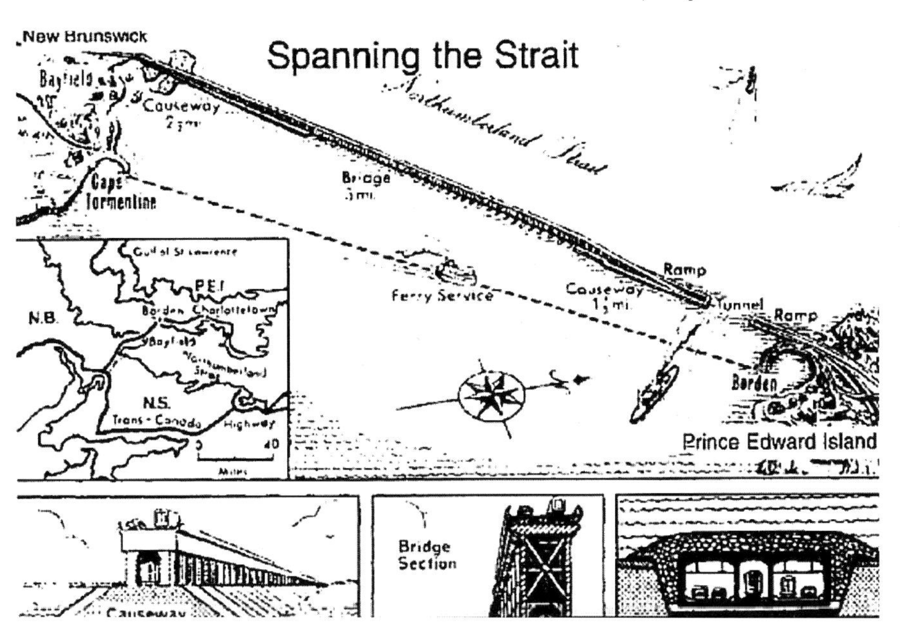

Figure 3.8 *On July 8, 1965, Prime Minister Pearson announced that the government would move forward on the construction of a combined causeway, tunnel, and bridge. Considerable work was done on the approach roads, but on March 5, 1969 the project was cancelled. On that date the federal government announced that instead of funding the causeway it would contribute $225 million over the next 15 years toward a* Comprehensive Development Plan *for Prince Edward Island.* [SCDI]

Proposal Call

On March 15, 1988 the final Generic Initial Environmental Evaluation document was released. It supported PWC's earlier conclusion that both a tunnel and a bridge were acceptable fixed-crossing alternatives, and on March 28 Stewart McInnis invited the seven prequalified developers to submit proposals by June 13. These proposals had to address the environmental risks identified in the GIEE, and demonstrate compliance with various performance, environmental protection, safety, and industrial benefit criteria. One of those seven organizations was Strait Crossing Inc., headed by a fellow named Paul Giannelia. Chapter 4 focuses on the 6-1/2 year effort by Giannelia and his associates to win the right to build the bridge.

Figure 3.9 *How do you make money from a plebiscite? By having thousands of buttons made, and selling them to both sides.*

CAN THE CAUSEWAY

P.E.I. joined Canada in 1873, and during the province's Centennial Year, 1973, some Islanders who were not so sure that Confederation had been an unmixed blessing formed a patriotic society called The Brothers and Sisters of Cornelius Howatt. (Howatt had been a member of the Legislative Assembly back in 1873, and had similar doubts.) The Brothers and Sisters engaged in all sorts of hilarious antics, including an attempt to put the causeway issue to rest forever.

On a December day in 1973, the Mayor of Borden shoveled a quantity of clay into a wheelbarrow as one of the Brothers read the following proclamation:

WHEREAS the construction of a Causeway to connect the Mainland with Prince Edward Island has long been a subject of much discussion, many broken promises, and innumerable false hopes;

And WHEREAS some ten years ago a group of naive college students did, in an act of impulsive and misguided enthusiasm, push a wheelbarrow load of good Island earth all the way from Charlottetown to Borden, where it was dumped over the cliff to mark the symbolic commencement of Causeway construction;

And WHEREAS it has now become self-evident to all thoughtful and patriotic Islanders that the actual completion of said Causeway would prove damaging to our economy, deleterious to our provincial autonomy, and destructive to our way of life;

THEREFORE be it known to all Islanders, that the Brothers and Sisters of Cornelius Howatt do hereby declare their determination to return said load of earth to the province's capital where it belongs, and, by the performance of said symbolic act, thereby, once and for all, now and for the future, effectively Can the Causeway.

To cap the symbolism the wheelbarrow was not only wheeled to Charlottetown, but following a torchlight parade to Province House the red clay from the wheelbarrow was, "with the aid of an old-fashioned home sealer, forever canned." The cans were then auctioned off, and their sale paid off "the debts accrued over the year by the Brothers and Sisters."

Adapted from the report in *Cornelius Howatt: Superstar*

Figure 3.10 |*Chambers/Halifax Chronicle-Herald*|

"Hold everything, boys—there's nothing over there but tourists, 'taters and Tories!"

November 12, 1965: Conservatives take every Federal seat in Prince Edward Island.

CHAPTER FOUR

EVOLUTION OF THE SOLUTION

To this point we've examined the transportation problem that Islanders faced over the years, and controversies over ways and means of solving that problem. Our tale now shifts from general to specific, and from problem to solution. The remaining chapters tell, stage by stage, the story of the bridge that now spans the Strait. This chapter focuses on the pre-contract period and the many ups, downs, and obstructions on the road to contract signing.

Beginnings

Paul Giannelia was in his late 20s when, in 1976, he was approached by the Ontario construction firm W.A. Stephenson about setting up an operation in Western Canada. Giannelia said yes, and in 1977 he started W.A. Stephenson (Western) Limited, with headquarters in Calgary. His partner of today, J. David Pirie, is the lawyer who incorporated that original company.

Calgary was booming in those days, but there was also lots of competition in the construction industry — especially building high rises, where Giannelia was most experienced. Looking around, Giannelia saw opportunities in public works projects. "There was less competition there," he recalls, "but the competition that did exist was very sophisticated. We had to come up with something different."

That something was the *unsolicited alternative proposal*. Giannelia may not have been the first person to have this idea, but it was a new approach in Western Canada. It works like this: A municipality decides it needs a new bridge. Traditionally, the city would hire a consulting engineer who would spend six months designing the structure and putting together a bid package of drawings and specifications. The city would then call for bids from construction companies. Each interested company would get copies of the drawings and specs, and after six or eight weeks would submit a price for building exactly what the drawings and specs called for. The city would review the bids, and the bidder with the lowest price would normally get the job.

There are many varieties of bridge design, many construction techniques, and many possible combinations of the two. Many different pairings of design and construction method would result in a functional and safe bridge. But among those pairings there would be major differences in what it would cost and the time required to build it. Giannelia sensed that if you brought design expertise and construction expertise together right at the beginning, you could often come up with a cost-saving or time-saving solution. This would please the customer and give the bidder a competitive advantage.

Since this was all brand new, Giannelia took a twofold approach. Just like all the other bidders, he would submit a bid on the bid-package design. But while some of his staff prepared that bid, other staff would explore alternatives. If a superior approach surfaced, the company would then prepare a second bid.

As Giannelia tells the story, "We would show up and say, 'Here is the bid as your engineer asked for it. And by the way, we think we can do a better job for you if we used this alternative design.' In doing this we ran head-on into a lot of problems. We were telling the designer and his client that we had a better idea than the designer had come up with. Nine out of ten of our bids got thrown out on the technicality that bid documents made no provision for this approach. My partners had some problems with continuing to do this, but I felt that we had to separate ourselves from the competition. In 1979 we started the first project for the City of Calgary on an alternate design that we had proposed. It was a very good job for us and a very good job for the city. We saved them about 25-30% over the low bidder, and that was the beginning."

Giannelia continued to refine this approach, in part by harnessing the computer. With the right computer and software you could design and price alternatives more quickly and with fewer people than you could by doing it with pencil and calculator. In the early 1980s, Giannelia's firm and the University of British Columbia set up a joint-venture company to develop computer systems for the construction industry. The fruits of this activity helped Giannelia bid with alternative designs.

The two-bid approach worked again in Vancouver in 1984 when one of Stephenson Construction's alternative proposals was accepted for the Cambie Street Bridge. The plus on this occasion was time. With the original bridge design, construction would not have been completed by the opening of Expo in April 1986. Stephenson's alternative bridge was completed four months before opening.

"The more we looked at it, the more it started to come together," said Giannelia. "The future is in what we today call 'one stop shopping.' You need to get from here to there, and to do that you need to come up with the environmental solutions, the technical solutions, the financial solutions, the managerial solutions."

With that sort of mindset, it's not surprising that Paul Giannelia got excited when he opened the *Globe and Mail* that day in May 1987 and read about the *Call For Expressions of Interest* concerning a tunnel or bridge across the Northumberland Strait. He had read about the unsolicited proposals made the year before, and had given the situation some thought. Now, here it was: a challenge worth addressing, and an official call to action.

Preparing to Bid on a Megaproject

As Giannelia tells it, "When I got back to Calgary I talked to my partners about putting our name

THE FIXED LINK

A BRIEF HISTORY...

©1992 Wayne Wright

Figure 4.1 *An eccentric view of the history of the bridge.* [Wright/The Journal-Pioneer]

forward. We also talked to some engineering people that we'd worked with before. We very quickly came to the conclusion that we couldn't beat the competition on total mass of past accomplishments or gigantic financial statements. They had all the big projects in the world in their picture books. *We had to have the best technical solution.* That was the only way that we could beat the kind of competition we were up against."

In June, a consortium led by Stephenson Construction expressed interest in the project. So did 11 other groups. In August, the Department of Public Works prequalified 7 of the 12, including the Stephenson group.

Figure 4.2 *Chambers' cartoon illustrates very nicely the waiting-for-the-ferry problem. Long waits caused aggravation for both visitors and Islanders. [Chambers/Halifax Chronicle Herald]*

With that hurdle over, Paul Giannelia started to put together a team of individuals and organizations to create that "best technical solution." The principle consultants were:

- Stanley Associates Engineering Ltd. of Calgary who led the engineering team and were the engineers of many other Stephenson projects;
- Speco Engineering, headed by Dr. Gamil Tadros, the designer of several bridges that Stephenson Group companies had built;
- Simpson Lester Goodrich Partnership (SLG), a structural engineering firm who had worked on several Stephenson projects including the Olympic Oval; and
- Leonhardt Andrä und Partner, a group of world class bridge experts from Germany.

Gamil Tadros recalls the team-building process: "We had to have someone with large-span experience, so we got Leonhardt. Dr. Walter Dilger of the University of Calgary had worked with Leonhardt, and we wanted him. We also needed subconsultants. We hired Allen Davenport from the University of Western Ontario who is known worldwide for his work with wind. We needed a soil expert, and we hired Golder Associates who stayed with the project to the end. And then we needed people who understood ice loads. We knew Jack Clark's work, so we hired his company, C-CORE. We hired people we were comfortable with."

"We also hired Bob Dewar, a Professor at the University of Calgary who specialized in ergonomics, the driveability aspect of bridge design. He's the one who told us to build the bridge with an s-curve in it. Our consortium was the new kid on the block, and I think that paying attention to even the ergonomic aspect of the project showed that we were very sincere in what we were doing."

SLG's Barry Lester recalls the day his firm joined the team: "I was talking to Paul one day in 1987. I will never forget the conversation. He told me that he was going after the bridge and was putting together a design team. And I said, 'Paul, how are you ever going to do that? The biggest things you've done are the Olympic skating oval that you just finished and the Cambie Street Bridge. How are you going to do a billion dollar bridge?' And Paul looked me and said, 'Heck, somebody's got to do it.' It's just a Sylvester-Stallone-type line. He wasn't phased by any of that; so what, a billion dollars. 'It's just scale,' he said. 'It's just more zeros.' Then he asked, 'Do you want to be on the team?' And I said, 'Sure.'"

Barry Lester recalls the team's first meeting at the Sandman Inn in Calgary: "Paul outlined what our objective was going to be in order to get this bridge. He started out by saying what we didn't have: We weren't from Eastern Canada. We didn't have maritime experience. We didn't have any strong political ties to the federal government or the provincial governments down there. He said it was a design-build-finance-operate project, and that we were going to have to have the best design, we were going to have to have the best construction methods, we were going to have to have the most attractive financing package, and we were going to have to have the most economical approach to maintenance and operations possible. And if we had all of those ingredients, it wouldn't matter where we came from or who we were, we'd get the job. And Paul said to us, 'Don't worry about the financing; we'll go get it. Don't worry about the operations; we'll take care of that once we've figured out what we're going to build. You guys go away and figure out the most economical way to build this bridge, and that will include both the design side and the construction side.'

"That was the challenge that he threw to the team at that time. Paul was always an integral member of the team, and I think that was one of the great strengths of this team. It wasn't a design team on one hand and a construction team on the other, it was very much design and construction together team, where the construction methods were an integral part of developing the design. So Paul said, 'We're going to review every possible kind of bridge in the world, and we're going to figure out what we're going to build here.'

"We started out by asking ourselves what kind of bridges we could build: girder bridges, suspension bridges, steel truss bridges, arch bridges, and we actually assigned roles to different members on the team, and asked different members of the team to prepare preliminary designs for a whole bunch of different kinds of bridges. We wanted a first cut at what these things would look like if you tried to stretch them over 13 kilometres."

The traditional way to build a long bridge is to construct it over the water in progressive fashion from one or both shores. This approach was soon called into question. Barry Lester again: "After the August prequalification — late October or early November of 1987 — Paul travelled to P.E.I. with Bob Butler of Stanley Associates. They arrived, and headed off to Borden in the midst of an incredible eastern blizzard. They got to the shore of Northumberland Strait, looked out, and couldn't see a thing. They came back to Calgary with a bunch of pictures that were just white, and at our next meeting Paul said, 'We're not building a damned thing out there on the water that we don't have to build on the water. We're going to build everything that we can possibly build on the shore. We're not doing piling operations out there on the water. We're not pouring concrete out there on the water. We're not even putting precast segments out there on the water if we can avoid it. We're going to make them all on shore and then take them out on the water.' In my opinion, that was the single most important decision that was made. It led directly to the way that this bridge was conceived and built."

Paul Giannelia's vision on the construction side of the bridge planning process was complemented by Gamil Tadros's vision on the design side. Dr. Tadros had received his Ph.D. from the University of Calgary, and during the 1980s worked closely with both Stanley Associates and Giannelia — designing several of the bridges that Giannelia's companies built. Tadros recalls the construction issue: "Paul's visions are very clear, and I always listened carefully to what he had to say. What I needed from him was not design suggestions, but how he *felt* about the construction process. What were his concerns, what were his fears, what bothered him when he pictured building it. In this case, what concerned him most was doing work in the Strait. By understanding that this was his fear, our focus moved to building it on land. Could we build the whole bridge on land and carry it out in one piece? Obviously not. Could we build a whole span and its foundation all in one piece and carry it out? No. So what were the largest pieces we could build, carry out, and set in place? How big could we make the pieces and have the contractor comfortable that he would be able to move and place them? This process of starting mentally with the whole bridge and cutting it into manageable pieces is how we came to major decisions about its design and how to build it. You cut the bridge into pieces that you knew you can handle."

Between November 1987 and May 1988 the team developed a conceptual bridge design and a scheme for building it. The details changed somewhat in the years that followed, but the preliminary design submitted to the federal government in June 1988 was very similar in appearance to the structure that is out in the Northumberland Strait today.

As is typical in the construction industry, W.A. Stephenson Construction (Western) Ltd. spawned subsidiary companies. One of these was Stephenson Construction International, a company that eventually became SCI Engineers and Constructors Inc., or SCI. Most of Giannelia's fixed-crossing competitors had names that identified them with the project: Borden Bridge Ltd., P.E.I. Bridge Ltd., P.E.I. Crossing Venturers, and Northumberland Bridge Corporation. What sort of a name would make sense for the fixed-crossing arm of Stephenson Construction? Then, one day, the answer came: Strait Crossing Inc. — SCI for short. What a saving on letterhead!

On June 13, Strait Crossing Inc. and the five other firms mentioned above submitted a total of 7 bridge proposals and 1 tunnel proposal.

Figure 4.3 *This artistic rendering of the bridge was part of Strait Crossing's bid, submitted to the federal government in June 1988.* [SCI]

During the summer these proposals were evaluated by 5 working groups set up by Public Works Canada (PWC). Each group focused on a separate aspect of the proposals:

- **Management** of the design, construction, operation, and maintenance of the fixed crossing;
- **Technical** aspects of the design, construction, operation, and maintenance;
- **Environment** impacts and their mitigation — how unavoidable impacts might be reduced or softened;
- **Canadian and Regional Benefits**;
- **Financing** including the combination of equity bonds, insurances, and letters of credit which secure the government of Canada and the developer against financial risk.

PWC was represented on these groups, but each also had its private sector specialists, representatives from the provincial governments involved, and other departments of the federal government. Each of the 8 proposals was subjected to a pass/fail evaluation in each of the five areas of concern. These were the unpriced "Stage II" technical proposals mentioned in Chapter 3. The intended result of this process was a short list of companies that would then submit final "Stage III" priced proposals.

An interesting feature of this process was that the developers were given two chances to meet the requirements. After the proposals were evaluated by the five groups, PWC called in the developers (individually) and discussed the shortcomings of their proposals with them. The developers then had three weeks to submit amended proposals.

On September 30 the final results were in. The government announced that only three of the original eight proposals met all the requirements of the proposal call. All three were for bridges. The tunnel proposal and four other bridge proposals were rejected. The three successful developers were

- Borden Bridge, for a combination segmental concrete and cable-stayed bridge;
- P.E.I. Bridge, for a concrete bridge with a cable-stayed section over the navigation channel; and
- Strait Crossing Inc., for a prestressed concrete bridge.

The original plan was to call for "Stage III" priced proposals on October 10, 1988. That didn't happen. It didn't happen, in fact, for several years. With a federal election in the offing, PWC decided to delay the call for Stage III priced proposals. The election came and went, and it was February of 1989 before Prime Minister Mulroney picked his new Minister of Public Works: Elmer MacKay. During the election campaign, concern for environmental issues had been high on the public agenda. This was particularly so in P.E.I. with regard to the fixed crossing. The federal Environmental Assessment and Review Process (EARP) required that if a proposed project might cause significant adverse environmental effects or engender major public concerns, the proposal had to be referred to the Minister of the Environment for an independent public review of the project's environmental and socio-economic effects. Elmer MacKay did this. A FEARO Panel was created, and for the next 15 months it studied the matter. (See **THE FEARO PANEL REVIEW** sidebar for details.)

Bidding and Negotiating

SCI would be submitting a bid and financial package at some point, and the company prepared for this eventuality in two ways. It worked out the financing, and it put together the bid team. During late 1991 and early 1992, J. David Pirie, SCI's Senior

Vice President, worked with Gordon Capital Corporation and CIBC Wood Gundy to develop a Real Rate Bond that would be attractive to the Canadian investment community and provide the funds needed to build the bridge. (For details, see the **FINANCING THE BRIDGE** sidebar.) He also developed the financial security package required by the government.

SCI also put together its bid team. Megaprojects are usually carried out by a consortium of several firms rather than by one firm alone. The three main reasons for this are risk sharing, the need for money, and the need for special expertise. Governments normally require performance bonds and other types of financial guarantees, and insist that project developers have a sizable amount of their own money at risk. All this takes deep corporate pockets, and normally no single construction firm would — on their own — want to accept all the risk for a project as large as this one.

SCI had become one of only three finalists in late 1988, and although SCI discussed partnership possibilities with various firms, it continued to pursue the project on its own throughout the environmental assessment period and the era of public debate. Then, on January 30, 1992, the Minister of Public Works announced that all three bridge proposals met the environmental requirements, and directed the companies to submit their bids and their financial and security packages. It was time to solidify the team, and potential partners now came courting in earnest. One of these was GTMI, the Canadian subsidiary of the French firm GTM Entrepose. Another was Morrison-Knudsen, an American firm. But who do you pick? Assuming that a potential partner could make the needed financial contribution, it was then a question of what else they brought to the partnership. The qualities of greatest value here were bridge-building know-how, and experience with large-project management. Both GTM and Morrison-Knudsen met these requirements. Decisions were made, and behind the Strait Crossing bid would be SCI, GTMI (Canada) Inc., and Northern Construction Company Ltd., the Canadian subsidiary of Morrison-Knudsen.

The federal government had always been clear about the method and amount it was prepared to pay for the bridge: no more than $35 million annually in March 1988 dollars — $42 million in 1992 dollars — for a period of 35 years. This was the government's estimate of the cost it would avoid by not having to provide a P.E.I./N.B. ferry service subsidy during that period. Paul Giannelia had always understood this, and the government figure had guided Strait Crossing's bridge design and financial planning activities from 1988 through 1992.

The three developers made their Stage III submissions, and on May 27, 1992 Public Works opened their bids. The Strait Crossing Inc. bid was $40.6 million — slightly less than the amount the government had established. The other two bids were well over that amount: P.E.I. Bridge Company Ltd.: $46.2 million, and Borden Bridge Company Ltd.: $64.2 million.

Unfortunately for Strait Crossing, there was a wrinkle. PWC's Financial Evaluation Committee ruled that Strait Crossing's "financial package" did not fully comply with the terms of the Proposal Call process. In fact, neither did the financial packages of the other two bidders. Strait Crossing never agreed that its bid didn't comply, but the Committee's ruling was what counted, and the bid process was closed.

What next? was the question on everyone's mind. On July 17 the answer came. PWC announced that it would start discussions with Strait Crossing Inc. — the low bidder — to see if a plan acceptable to both parties could be worked out. Discussions were held through the summer and fall. Then, on December 2, 1992 in an economic statement to Parliament, Minister of Finance Don Mazankowski announced publicly that Strait Crossing Inc. had been selected for the project. The company would be allowed to proceed with engineering and environmental work, and negotiations would begin between the government and the company to finalize the terms of the contracts.

The federal government was not the only government with a say in whether or not the project went ahead. The undersea land upon which the bridge would be built had to be deeded by P.E.I. and New Brunswick — and then there were Joe Ghiz's 10 Commandments. Three-way governmental negotiations took place in 1991 and 1992, and eventually resulted in the tripartite agreement of December 16, 1992. This document addressed all of Premier Ghiz's concerns, and in some respects gave P.E.I. more than the Premier had originally asked for. For example, Ghiz had required that the federal government carry out studies on Borden redevelopment. In the tripartite agreement, the federal government agreed to supply at least $10 million for actual development activities in Borden. For his part, Premier McKenna of New Brunswick was assured that the headquarters of Marine Atlantic would not be moved out of his province, and he got money for both Cape Tormentine redevelopment and approach highways.

Shortly thereafter, on December 22, 1992, Strait Crossing released their Environmental Management Plan (or EMP) in preparation for public review — as required by the FEARO Panel report. (See sidebars.)

Figure 4.4 |*Canon Robert C. Tuck*|

Days in Court

A final agreement now seemed near, and Strait Crossing was counting on getting a lot done during the 1993 construction season. It was not to be. For one thing, the negotiations were more complex and prolonged than anyone had anticipated. The project was breaking new ground in many areas. There were no precedents for some of the agreements that were needed, and all parties were trying to minimize their exposure to unfamiliar risks.

In addition, two Federal court actions were brought against the Government of Canada by the Island group who opposed the project, Friends of the Island, and this put another brake on the process. In their first motion, filed December 6, 1992, Friends of the Island claimed that the Minister of Public Works had failed to follow certain procedures under Section 12 of the Environmental Assessment Review Process Guidelines Order, and asked that the Minister of the Environment appoint an Environmental Assessment Panel to review the SCI proposal. They also asked the court to declare that a fixed link would contravene the 1873 terms of union, and asked it to prohibit the government of Canada from entering into any fixed-link agreement.

The case was heard before Madame Justice Barbara Reed in March 1993. In her decision later that month she agreed that the Minister of Public Works had failed to comply with the requirement to make a Section 12 decision "respecting the specific SCI proposal" but did not say that the SCI proposal had to be reviewed by an EARP Panel. She agreed that discontinuing the ferry service without a constitutional amendment would be a breach of the *Terms of Union*, and ordered that "the Government of Canada shall not make any irrevocable decision relating to the specific SCI proposal until after a section 12 decision is made . . ."

The FEARO panel had worked with a generic bridge design. Strait Crossing, on the other hand, always worked with its own specific design. By this time SCI had already analyzed many bridge/environment interactions for its bridge. Thus, producing a *Specific Environmental Evaluation* (SEE) was not difficult or particularly time consuming. There was already a massive amount of information on hand in the many consultant studies already done. There were some highly refined modeling tools. And from past activities such as the GIEE, the FEARO Panel, the draft EMP, written statements, and public meetings, SCI had a very complete idea of what the environmental experts and the general public felt should be considered. These prior activities had already defined issues, concerns, and components of the eco- and socio-economic systems that the public clearly valued.

The authors of the SEE addressed each concern that had been identified, and assessed the impact of the SCI bridge on each "valued ecosystem component" and "valued socio-economic component" relevant to that concern. Fifteen Valued Ecosystem Components were identified, ranging from the Salt Marsh Aster (an endangered plant species), to groundwater resources, to plankton productivity. The twenty-one Valued Socio-Economic Components included the commercial fisheries, land use and development, quality of life, and the Marine Atlantic ferry workers.

An impact's significance depended on its extent, duration, and magnitude — as well as the component's sensitivity to, and ability to recover from, the impact. Those doing the evaluation developed clear definitions of impact significance for specific valued components, and presented those definitions in the report.

The SEE report was released on April 22, 1993 and sent to the Minister of Public Works for his Section 12 assessment. Opponents of the bridge cried foul, because they couldn't imagine a thorough evaluation being done so quickly. In fact, the evaluation had been going on since 1988. PWC's March 1988 Proposal Call document devoted 104 pages to the "Requirements for Environmental Assessment, Review and Planning." Any contractor who focused only on steel and concrete and didn't give a damn about the environment was not going to get this job. To be the successful bidder, the developer would have to *embrace* the environmental

Figure 4.5 *Paul Giannelia, the Project Director of Strait Crossing Joint Venture, became a well known figure in Prince Edward Island.* [*Wright/The Journal-Pioneer*]

aspect of the project, and devote the same sort of intelligence, care, and commitment to environmental matters as it devoted to design or financing. And this is exactly what SCI had done — as demonstrated in their EMP.

Following the SEE submission, four public meetings were held to present the report's findings to the public and provide an opportunity for public discussion. On May 13 Elmer MacKay, the Minister of Public Works, announced that his department had completed the court-ordered assessment and had found that any environmental impact of the SCI bridge would be either insignificant or could be overcome by known technology. He had 30 days to decide whether or not to hold public hearings. On June 24 the Minister announced that "the issues raised throughout the process ... have not changed" and "Public Works Canada is satisfied that all issues raised have been addressed. Hence, public concern about the proposal is not such that a public review is desirable."

Elmer MacKay's May 13 announcement prompted Friends of the Island to go back to Federal Court on June 7 with a motion to quash the Minister's decision and force the appointment of an Environmental Assessment Panel. The motion also asked for a declaration that it would be unconstitutional for SCI and the government to enter into an agreement to build the bridge.

On August 12, Mr. Justice Bud Cullen denied that motion. In his 88-page decision he said "I believe the very extensive assessment done here was far more than the minimum called for under the guideline order. One has to be impressed with the extent of the work done ..." Regarding whether it had been done properly, Mr. Justice Cullen said: "The SEE runs through nine chapters and then in Chapter 10 reaches its conclusion which in my view is unassailable and certainly the applicant [Friends of the Island] has been unable to mount any real attack on either the methodology or the conclusion."

On the constitutional issue, Mr. Justice Cullen declared that "there is nothing inherently unconstitutional about building a bridge or entering into a contract to build a bridge. . . . What is unconstitutional is discontinuing the ferry service without an amendment to the *Prince Edward Island Terms of Union*." He saw no problem with "PWC and SCI entering into an agreement to build a bridge, if such an agreement contains an undertaking to amend the Constitution prior to discontinuance of the ferry service."

The Home Stretch

Mr. Justice Cullen's decision brought a huge feeling of relief to those who wanted the project to go ahead, and it put energy into the final negotiating and document drafting process. For the next 50 days officials of Strait Crossing and Public Works, together with teams of lawyers for both sides, put together the several hundred documents that were needed to close the deal.

Back in June of 1993 the House of Commons had passed Bill C-110, the enabling legislation which allowed the Minister of Public Works to build a fixed crossing across the Strait, to collect tolls for the use of the crossing, and to pay an annual inflation-indexed subsidy of up to $42 million in 1992 dollars for a period of 35 years.

Also in June, P.E.I.'s new Premier, Catherine Callbeck, had entered into an agreement with the federal government concerning a constitutional amendment. Premier Ghiz had opposed any change to the constitution, but Madame Justice Reed's decision convinced many that amending the

Figure 4.6 *A humorous look at the court action relating to the bridge.* [Wright/The Journal-Pioneer]

Terms of Union was probably essential, and definitely prudent. Premier Callbeck and her government agreed to the addition of the following wording:

> That a fixed crossing joining the Island to the mainland may be substituted for the steam service referred to in this Schedule;
>
> That, for greater certainty, nothing in this Schedule prevents the imposition of tolls for the use of such a fixed crossing between the Island and the mainland, or the private operation of such a crossing."

These passages were added to the Constitution on April 15, 1994.

In return for agreeing to this change, the Premier got a concession regarding tolls. All along it had been clear that annual bridge toll increases would be limited to 75% of the annual rise in the Consumer Price Index. This meant that tolls, in real terms, would gradually go down during the first 35 years of bridge operation. But what would the tolls be *after* those first 35 years? The federal government agreed that when this 35 year period ended, bridge tolls would be limited to an amount covering only the operating and maintenance costs of the bridge, not the cost of replacing it.

With a mountain of documents to be signed, closing was a spread out affair. Hundreds were signed in Toronto on October 7; the final few were signed in Charlottetown on October 8. It had taken 77 months — six years and five months — for SCI to progress from that first request for expressions of interest to actually signing a deal to build the bridge.

Looking back, what did Paul Giannelia think of it all? "The ongoing challenge for me was dealing with all the negativity. Today, it surprises me that there was so much of it, and that it took so long to die out. The biggest challenge, at the end of the day, was to make believers of the bystanders, the people in the middle, and people on your own team.

"Back in 1987 I estimated the cost of bidding this job at $243,000. In the end it was a hundred times that — $20 or $25 million. Once we started spending that kind of money it was easy to stay focused on winning it. But the overriding thing about the project that gave me the confidence to push on was that it made economic sense to build it. I knew that it would eventually get built. It wasn't an *if*; it was always a *when*. The arithmetic told you that. It made economic sense for the taxpayer: after 35 years the ferry subsidy was no longer needed. And there were other economic pluses such as getting goods to market more easily. It made economic sense, and that's what kept us going."

Figure 4.7 *In Charlottetown on October 8,1993, Paul Giannelia of Strait Crossing (left), Bernard Valcourt, Federal Human Resources Minister (centre), and Peter McCreath, Federal Veterans Affairs Minister (right) signed the historic deal to build a bridge across the Northumberland Strait.* [SCDI]

THE FEARO PANEL REVIEW

On March 8, 1989 — before selecting one winner from the three proposals then on the table — the Minister of Public Works, Elmer MacKay, referred the whole fixed crossing matter to the Minister of Environment for an Environmental Assessment Review Process (EARP) review.

The matter was handled by the Federal Environmental Assessment Review Office, and the panel doing the assessment soon came to be called "the FEARO Panel." Its mandate was broad. It was to review the effects and risks of project construction, operation, and maintenance in 13 areas, including

- marine and terrestrial plants and animals;
- changes in tides, currents, and inshore dynamics;
- changes in ice climate, including formation and breakup;
- physical interference with commercial fishing activities; and
- socio-economic effects within the region and on particular communities, e.g. regional industrial benefits, and effects on ferry workers and fishers.

To provide information for both the Panel and the public, PWC released a *Bridge Concept Assessment* document. At this point there were still three competing developers and their three bridge designs. To maintain confidentiality, PWC could not give the actual designs to the FEARO panel. Thus, the PWC document presented a hypothetical design to the panel: 75 to 80 support piers in the water, a 175 metre span between piers, 10% blockage of the Strait by the piers, and 300,000 cubic metres of dredging.

Between June 19 and June 29, 1989 the panel held 12 meetings at locations in P.E.I., New Brunswick, and Nova Scotia, and heard 51 presentations. Gilles Theriault, former President of the Maritime Fishermen's Union and a panel member recalls, "We were not expert in every field, and there were a lot of things that were very complicated. But we had the budget to bring in anyone from anywhere in the world to help guide us through the issues." Six of these advisors are mentioned in their final report: experts on risk assessment; fisheries and marine biology; tunnelling; ice regime; mitigation and compensation; and socio-economic impacts.

At the end of August the panel requested more information from PWC, who responded on December 15 with a *Supplement to the Bridge Concept Assessment*. The public was invited to submit written comments "on the sufficiency of the additional information." After reviewing the *Supplement* and thirty submissions from the public, the Panel called another round of public hearings. Over 1500 people attended 21 sessions, and the panel heard 150 presentations.

On August 15, 1990 — after 15 months of work — the FEARO panel released its final report. Most people were expecting a YES, a NO, or a YES BUT. Some people read the report and concluded that it was saying a firm NO to a bridge. Others read it and concluded that it was a conditional NO, a NO BUT. They saw the report saying NO UNLESS OUR OBJECTIONS HAVE BEEN SATISFIED. Asked whether the panel's report was intended to kill the project completely, Gilles Theriault replied, "Maybe in some panelists' view, but not in others. I can't speak for the others, but in my case I felt that the type of report that we had written left the door open for further assessment of the key issues."

And further assessment is exactly what those issues received. Public Works stepped back from the report's overall conclusions and focused on the specific objections which led to the conclusions. It then set about to see if all the Panel's objections could be met. On November 21, 1990 PWC responded to the Panel's report. This response pointed out that the Panel had studied PWC's *generic* bridge concept, and that some of the specific designs put forward by the developers "already meet or exceed the standards recommended." The PWC document then reviewed each issue raised by the Panel and responded to it. In those responses it acknowledged the legitimacy of the Panel's concerns, and it outlined ways and means of addressing them.

A major concern of the Panel was that a bridge would delay the date of ice-out by a week or two. This, in its judgment, would be unacceptable to the fishery and the coastal microclimate on which local agriculture depends. But the FEARO report also suggested that a maximum ice-out delay of two days in any year in a 100 year period would be acceptable. Here, the PWC response was to accept the Panel's 2-day ice-out delay criteria, and to appoint an independent committee of renowned ice experts to review and refine the existing model of ice-clearing/ice-jamming behaviour. What was desired was a model that would accurately predict the influence of a bridge on ice-out delay, and relate the amount of delay to bridge design parameters such as pier diameter and pier spacing.

The Ice Committee, as it was called, was formed in January of 1991. In April it published a report which concluded that "a bridge which meets the criteria for ice delay established by the FEARO panel can be installed across the Northumberland Strait." The committee then applied the refined ice model to the three specific bridge designs. In their report of the results, the Ice Committee predicted "that all the three bridge designs submitted will meet the 2-day criteria with a comfortable margin."

On May 9 the Minister of Public Works announced that the three developers would be invited to resubmit their proposals and demonstrate how each design meets the environmental requirements drawn from the FEARO panel report. An Environmental Committee composed of representatives from PWC, Environment Canada, Fisheries and Oceans, and the environment departments of the Maritime provinces reviewed the proposals, and concluded that they met the FEARO panel requirements.

FINANCING THE BRIDGE

The final agreement between Strait Crossing Development Inc. and the federal government stipulated that the annual subsidy will be $41.9 million in 1992 dollars, adjusted each year in accord with the Consumer Price Index. Discounting the future value of this stream of yearly payments, fully indexed to inflation and guaranteed by the government of Canada, made it possible to issue Bonds yielding a guaranteed "Real Rate" of return in the Canadian Capital Markets.

The Real Rate Bond mimicked an existing federal debt instrument, and its use was researched and developed by SCI in its proposals in the late 1980s. A company called Strait Crossing Finance Inc. was set up as a New Brunswick crown corporation to receive the annual indexed payments from the federal government and pass them on to a Trustee for the Real Rate Bond holders. This structure resulted in the Bonds not being exposed to any credit or litigation risk other than the credit risk of the Government of Canada. Consequently, the bonds received an AAA credit rating from Moody's and Standard & Poors, and pricing comparable to the Real Rate Bonds offered directly by the government of Canada.

This offering raised approximately $660 million, and it is this money plus the interest it earned during the construction period that would pay bridge construction costs. It was held in trust and withdrawn upon approval of the Independent Engineer who monitored design and construction activities.

To secure the government against failure to complete the project, the developer provided an extensive security package consisting of parent company guarantees, a $200 million Performance Bond, a $20 million Labour and Material Payment Bond, and a $35 million Defects Assurance Bond. To secure it against cost overruns, the government required a separate Letter of Credit for $73 million — 10% of the estimated direct construction cost of $730 million.

Return on investment and risk for the project partners will come from the toll revenue stream after bridge operating and maintenance expenses have been paid.

STRAIT CROSSING'S ENVIRONMENTAL MANAGEMENT PLAN

The Stage III bidding process required that the successful developer prepare an Environmental Management Plan (EMP) and release it for public review. This EMP document had to specify how all the environmental aspects of the project would be managed (both during construction and for 35 years afterward) to minimize and mitigate effects on the environment. This EMP would have to be accepted by the government's Environment Committee before there could be a financial closing and agreement signing.

Strait Crossing had been working on its EMP throughout the last half of 1992, and on December 22 released a draft for review by the public and the Environment Committee. Between January 7 and 21, 1993, Strait Crossing conducted 9 information sessions concerning this draft Environmental Management Plan.

This time the meetings were hosted by Strait Crossing, and the four people who did the research, made the presentations, and fielded the questions from the audience were Strait Crossing people. Paul Giannelia and his team had prepared for this. "By this time we knew what was missing in the earlier public meetings," said Giannelia. "And what was missing was *answers*. We now knew what the answers were, and we entered this round of meetings with a lot of confidence. The FEARO report put in black and white what the real issues were for the community. And this became the cornerstone around which we built our presentations. We took it chapter and verse. That's the statement; what's the answer? That's the statement; what's the answer? That's the statement; what's the answer? We also took all the years of newspaper clippings — everything everyone had said against the bridge — and made sure we dealt with it in our presentation. It was, however, a very time-consuming and costly affair."

Based on comments from the public in these meetings, and comments from the Environment Committee, Strait Crossing revised the draft EMP. The revised version was accepted by the Committee, and Strait Crossing began collecting baseline or "pre-bridge" data as the first part of a life-of-project Environmental Effects Monitoring (EEM) program.

Figure 4.8 *Extensive surveys of ice condition in the Northumberland Strait provides essential information for ongoing studies of the environmental impact of the bridge.* [PWGSC]

CHAPTER FIVE

HITTING THE GROUND RUNNING! 1993-94

During 1993, the public spotlight was on public meetings, court battles, bond issues, and final contract negotiations. At the same time, much was happening behind the scenes so that the project could start as soon the ink was dry on the contracts.

Forming the Team

The firms who put up the security, took the risks, and committed themselves to build the bridge were:

- Strait Crossing Inc.
- GTMI (Canada) Inc., and
- Northern Construction Company Limited

GTMI (Canada) Inc. is a subsidiary of the French firm GTM-Entrepose. Based in Montreal, the company has deep routes there. GTMI's main subsidiary and predecessor company is the JANIN Group, named for it's founder Alban Janin who was manager of the Civil Work Department of the City of Montreal early in this century. GTMI / JANIN projects include the Pierre LaPorte suspension bridge in Quebec City, the Congress Convention Hall in Montreal, the Des Jardin Complex, some large hydro-electric projects, and the Hibernia Gravity Base Structure. Recent GTM-Entrepose projects include a 5 km bridge linking England and Wales, viaduct and mass transit projects in Singapore, and the Hong Kong international airport.

GTM's Gilles deMaublanc commented on the GTMI connection and the reasons for GTM's involvement: "One reason was that this was a Build-Operate-Transfer project, and the BOT concept is very attractive to us. The second reason was that GTM has had a presence in Canada of more than 60 years under the name JANIN — a company that has been involved in big bridge and dam construction. So we are very Canadian. Third, the company had extensive experience in large size projects and there was a need for this special experience and expertise."

The other partner, Northern Construction Company Limited, was the Canadian subsidiary of Morrison-Knudsen Company — an international engineering, construction, mining and environmental company based in Boise, Idaho. The company had extensive bridge experience, and was a member of the joint venture team that built the Sunshine Skyway over the Tampa Bay in Florida. Unfortunately, the company ran into financial difficulties in 1995, and withdrew from the Northumberland Strait Bridge Project in 1996. Their 36% share in the joint venture was purchased by the remaining partners.

During 1993, Strait Crossing became involved with a company that would soon become a fourth partner in the joint venture: Ballast Nedam Canada Limited, a Dutch firm and one of the world's major construction groups. Ballast Nedam also had extensive bridge building experience. Back in the early '80s they built a 25-km causeway between Saudi Arabia and Bahrain, and they played a major role in the design and construction of the 7 km long West Bridge of the Storebælt project in Denmark.

The Storebælt project had a special relevance to the Northumberland Strait Bridge Project. Leonhardt Andrä (the German bridge design company) was on the Strait Crossing design/construction team back in 1987-88 when the build-it-on-land approach was worked out. Years later, Leonhardt became involved with the West Bridge project. There were certain similarities between the P.E.I. and Danish situations, and the West Bridge consortium decided to build their bridge in a manner similar to the one that had been worked out for building the Northumberland Strait bridge. Large pieces of bridge would be built on land, then transported by water and placed.

As we know, the Canadian project was delayed for years, but the West Bridge project was not. In fact, before construction of the Northumberland Strait bridge began, the West Bridge had been completed. This was a plus for Strait Crossing. As Kevin Pytyck, Strait Crossing's Manager of Contract Requirements, explained, "Storebælt took a concept we proposed and implemented it, and this was a reconfirmation of our ideas. It also taught us a lot about the challenges we were going to face. Dredging. Moving things around the construction yard."

Storebælt had something else for Strait Crossing: a large twin-hulled floating crane called the *HLV Svanen* that had been designed specifically to carry and place large bridge components on the West Bridge project. It had done its job well, and was now for sale. In late 1992, a group from Strait Crossing visited the West Bridge project, and the next summer SCDI acquired an option to purchase the *Svanen* from Ballast Nedam. The two firms also discussed the possibility of Ballast Nedam joining the Strait Crossing consortium, and in 1994 Ballast Nedam did join.

The *Svanen*, as it existed at that time, would not meet the needs of the Canadian project. The largest of the planned bridge components was heavier than the *Svanen* could handle, and it had to be lifted to a greater height. To be useful here, the *Svanen* had to be extensively modified, and that was eventually done in France in 1994-95. (Detailed information about the *HLV Svanen* appears in Chapter 6.)

In the post-contract environment of late 1993 and beyond, the joint venture partners participated directly in the project through two organizations:

- Strait Crossing Development Inc. (SCDI), the developer with whom the federal government had its agreements, and
- Strait Crossing Joint Venture (SCJV), the construction consortium that built the bridge under a construction contract with SCDI.

Partner involvement also extends to bridge operation. SCDI has agreements with Strait Crossing Bridge Limited, the wholly-owned subsidiary company that will operate the bridge for 35 years.

In the fall of 1993 the final design team began to take shape. SLG Stanley of Calgary, Alberta, a participant in the project design team since 1988, had joined forces with J. Muller International (JMI) of San Diego, California to form JMI/Stanley Joint Venture Inc. (JMS). JMI had previously worked for SCI, undertaking alternate designs and value engineering on several major bridge projects including the H3 Windward Viaduct in Oahu, Hawaii. These two companies formed an effective team, with SLG Stanley's extensive knowledge of the project since its inception and JMI's vast experience in the design and engineering of precast segmental bridges.

The scope of JMS's work included the complete detailed design of the structure, an internal design check, producing more than 10,000 design and shop drawings, on-site quality monitoring and technical support, writing the maintenance manual, and coordinating design criteria and specialty design consultants. The management of all design and engineering activities, including the coordination and administration of JMS's contract, was the responsibility of the SCJV Design/Engineering Management Department.

Not directly involved in the final design work, but auditing and critiquing it, was SCJV's Technical Review Committee. SCI's Gamil Tadros chaired the SCJV Committee, and its other members — Jacques Combault of GTM, George Fortinos representing Morrison Knudsen, and Willem Bilderbeek of Ballast Nedam — represented these firms at the most senior engineering level. They reviewed the evolving design on behalf of the developer. The North Vancouver firm of Buckland and Taylor was the Independent Engineer — the project overseer for the federal government. The Independent Engineer performed an independent design check, monitored construction quality, and controlled the release of funds.

During the summer of 1993 SCI negotiated with labour unions. Some years before, SCI had met with the P.E.I. and N.B. Building Trades Councils and was satisfied that there was sufficient skilled labour available from the area. Thus, in its proposal to Public Works, SCI had committed itself to supplying 96% of project employment from Atlantic Canada.

Paul Giannelia went into the 1993 labour negotiations with a clear idea of what he wanted. For one thing, he wanted to keep things as simple as possible. The bridge was physically large, but it had a concrete-and-steel sameness from shore to shore. Compared with many megaprojects, the range of skills needed was small. Giannelia had identified five trades that he considered essential, and he wanted agreements with those trades only. Trades that were not part of this group of five were understandably upset, but Giannelia kept pointing out that "It's just a bridge," and that not all trades were needed. He repeated that line so frequently that it became a slogan, and the day the agreement was signed all the union reps and negotiators were given Strait Crossing sweatshirts with "It's just a bridge" on them.

The five unions and Strait Crossing signed the project agreement on September 17. This agreement specified wages and various other conditions, an annual wage increase based on the inflation rate, and no strikes or lockouts for the duration of the project. As Kevin Pytyck described the situation, "For Strait Crossing, securing a labour agreement which established a positive proactive relationship with principle trade unions was fundamental to proceeding with the project. And in contrast with the labour agreements for many projects, its content was revolutionary."

Figure 5.1 *Making up the Confederation Bridge's overall length of 12.9 km is a 1.3 km section of approach bridge over shallow water at the west (New Brunswick) end, 11.0 km of main bridge over deep water, and a 0.6 km section of approach bridge over shallow water at the east (Prince Edward Island) end. The approach bridge spans from pier to pier are typically 93 metres long, and there are 21 of those. The main bridge spans are a quarter of a kilometre long (250 metres), and there are 43 of those. There are 44 main bridge piers, numbered P1 to P44 starting at the P.E.I. end.* [SCDI]

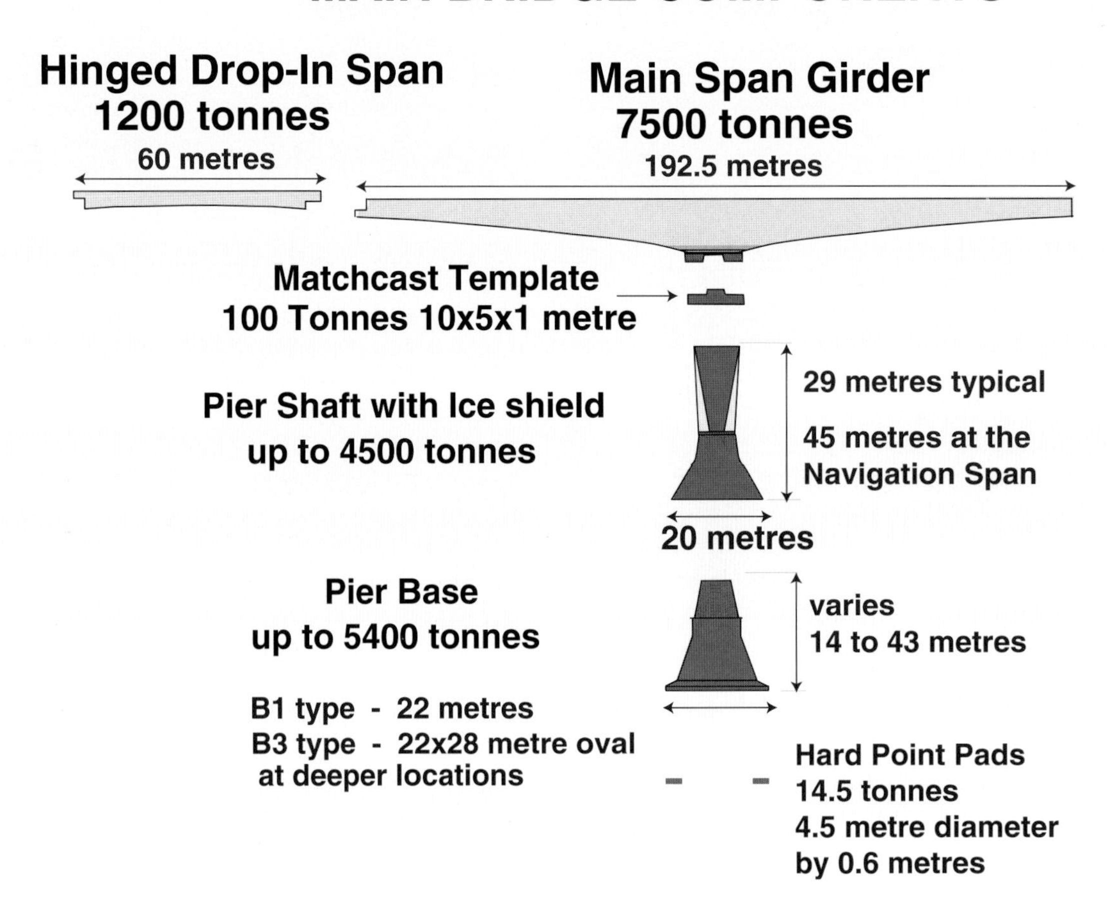

Figure 5.2 [PWGSC *with* C.M.]

An Overview of the Task

Getting the contract signed, teams formed, and the work force lined up were important 1993 milestones. Acquiring suitable places to build bridge components was another. To grasp what "suitable" might mean, we need more details about the nature of the task.

The bridge profile drawing of Figure 5.1 gives a rough sense of size and quantity. Making up the overall length of 12.9 km is a

- 1.3 km section of approach bridge over shallow water at the west (N.B.) end,
- 11.0 km of main bridge over deep water, and a
- 0.6 km section of approach bridge over shallow water at the east (P.E.I.) end.

The approach bridge spans from pier to pier are typically 93 metres long, and there are 21 of those. The main bridge spans are a quarter of a kilometre long (250 metres), and there are 43 of those. There are 44 main bridge piers, numbered P1 to P44 starting at the P.E.I. end. Needless to say, that's a lot of bridge.

We learned in Chapter 4 that the 1987-88 design/construction team decided that as much of the bridge as possible should be fabricated on land, and that the number of different components should be kept to a minimum. In 1993 the SCJV design and construction teams returned to this problem. They had to make final decisions about how to divide up the bridge, and what equipment to obtain for lifting, transporting, and placing the pieces. In the end, everything had to make sense from both a design perspective and a construction perspective.

Figures 5.2 and 5.3 show the set of pre-built main bridge sections that the design team decided upon, and how those sections fitted together. All the components were to be made of steel-reinforced high-strength concrete, further strengthened and linked together with post-tensioning tendons of high-tensile-strength steel. All but the smallest components were hollow. This made structural sense (more strength per unit weight), and it also provided a passageway within the bridge, from one end to the other, through which utility cables could be run. The heaviest component, the main girder, would weigh slightly less than the maximum rated lifting capacity of the modified *Svanen*.

If the water had been deep all the way to the two shores, the main bridge design would have been used from shore to shore. In reality, the water near each shore is too shallow to accommodate the *Svanen* and other main bridge marine vessels. Thus, a quite different *approach bridge* design was needed at the bridge ends, together with different construction methods. Like the main bridge, the two approach bridges also utilized precast elements, but they were much smaller and lighter than the main

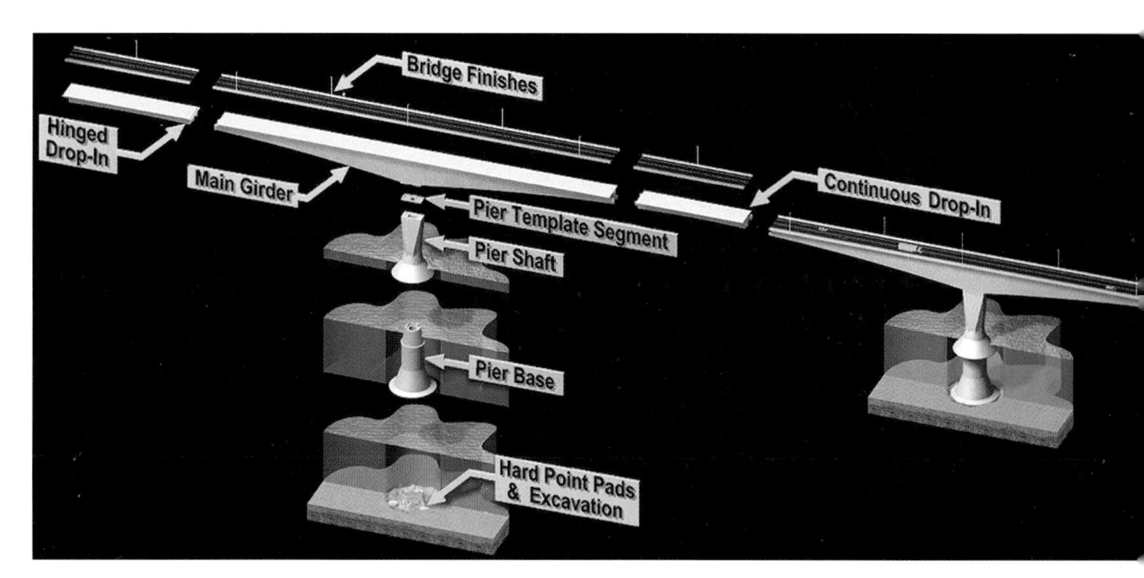

Figure 5.3 |SCDI|

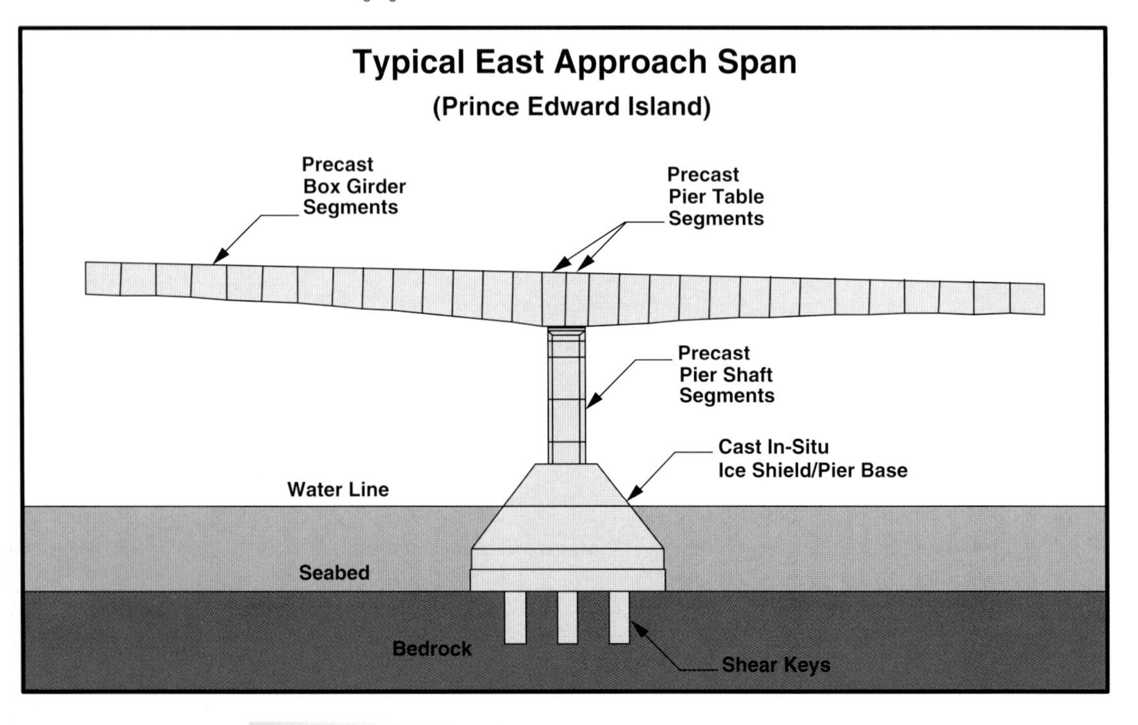

Figure 5.4 *The heaviest approach bridge element, at about 90 tonnes, is the Pier Table Segment; the other elements weigh 60 to 65 tonnes. Post-tensioning steel tendons inside the box-like structures pull the segments together, making a complete one-piece girder or pier shaft. A tight fit between segments was assured by "matchcasting" adjacent segments. With this technique, a previously cast section served as part of the formwork when the adjacent section was cast.* [SCDI]

←

Figure 5.5 *A typical approach bridge girder segment. It took 28 of these segments and two Pier Table Segments to make one 93-metre approach bridge span.* [PWGSC]

bridge components. Figure 5.4 shows a typical east-end approach bridge span. It used three types of precast components:

- **Girder Segments** — 28 per 93-metre span.
- **Pier Shaft Segments** — from 2 to 10 depending on the height above water.
- **Pier Table Segments** — 2 per span.

The heaviest component, at about 90 tonnes, was the Pier Table Segment; the other components weighed 60 to 65 tonnes. (A tonne is a metric ton. 1 tonne = 1000 kg = 2205 pounds.) Again, post-tensioning steel tendons inside the box-like structures pulled the segments together, making a complete one-piece girder or pier shaft. A tight fit between segments was assured by "matchcasting" adjacent segments. With this technique, a previously cast section served as part of the formwork when the adjacent section was cast.

Figure 5.5 shows one of these "small" approach bridge girder segments. Both these segments and the main girders are 11 metres wide — wide enough to accommodate two lanes of traffic and two breakdown lanes. (Additional main bridge and approach bridge details appear in the **DESIGN SPECIFICATIONS** sidebar.)

Constructing an Open-Air Factory

The 165-acre Amherst Point farm that Strait Crossing purchased from John L. Read was an ideal site for the main bridge staging facility. (See the sidebar: **SELLING THE FARM**.) But it took time and about $50 million to transform the first into the second. While the Reads still owned the property, surveyors carefully documented the lay of the land, and the engineers figured out how much earth had to be moved. An environmental team did their assessment, and knew exactly where the siltation barriers had to be placed in order to keep soil from entering the adjacent marsh. As soon as the Reads left, outbuildings were removed and McNamara Construction of St. John's Newfoundland brought their earth-moving equipment onto the property. They had a $4.5 million contract to prepare the site.

Strait Crossing's Steve Marko supervised construction of the 60 hectare (150 acre) fabrication yard: "McNamara arrived on November 1, 1993,

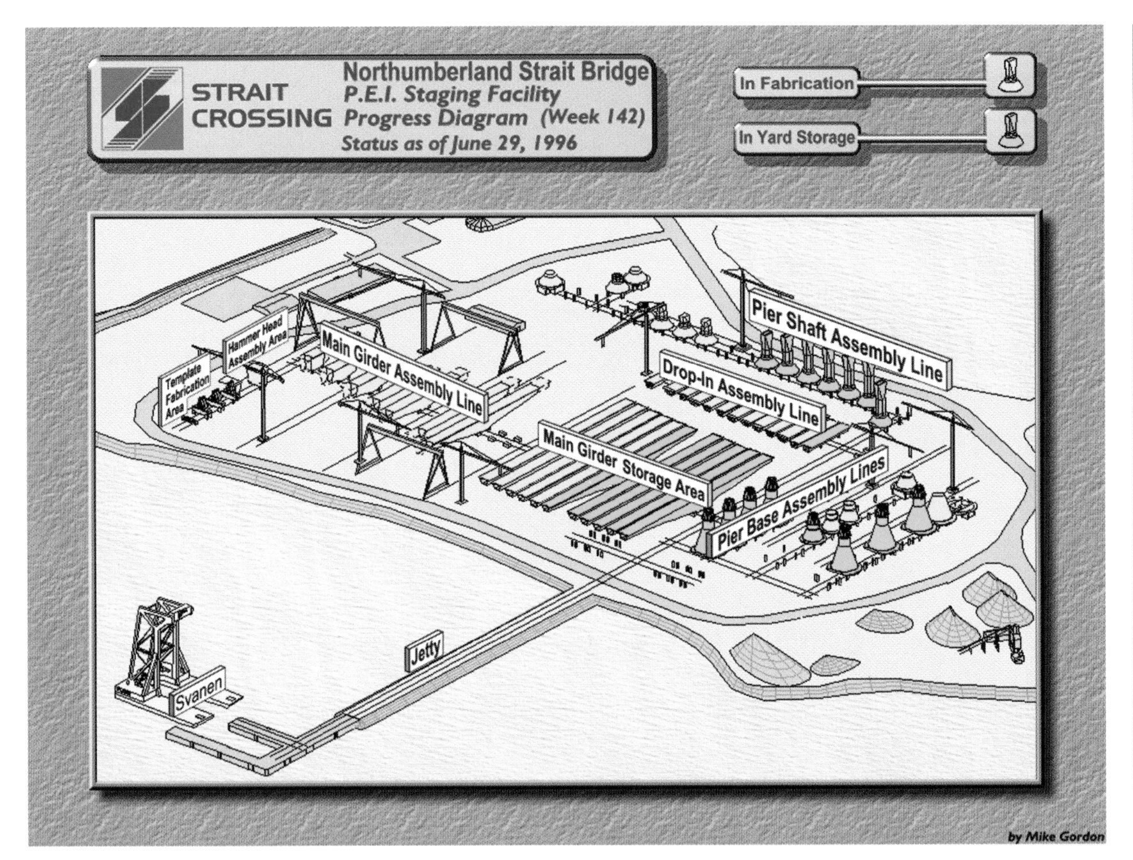

Figure 5.6 *An artist's conception drawing of the P.E.I. Staging Facility where components for the main bridge were pre-fabricated.* [SCDI]

Figure 5.7 *This September 1994 photo shows some of the concrete pillars upon which the large main bridge components will rest, and some of the 6000 metres of skidway upon which the Huisman sledges will slide as they move those components from place to place in the P.E.I. Staging Facility.* [Boily]

and we got right to work. Our first duty was to construct the pad for the 25,000 square foot office building. We then drilled for water supply. Over the winter months we moved in excess of 1,000,000 cubic metres of material to create a level site 5 metres above mean sea level, all with internal drainage." Another important task was creating a jetty about 500 metres long out into the Strait so

1996 photos of a Huisman at work moving one 7500 tonne main girder — equivalent in weight to 5500 compact automobiles. [Boily]

Figure 5.8

Figure 5.9

that the *Svanen* could remain in deep water and still pick up components. An artist's conception of the completed facility is shown in Figure 5.6 (see page 61). It is variously referred to as the *P.E.I. Staging Facility*, the *Borden Fabrication Yard*, and the *Borden Casting Yard*.

To understand why the Yard had to be level it is necessary to understand how components would be moved from place to place on one of two Huisman sledge transporters named the *Turtle* and the *Lobster*. All the large main bridge components — main girders, drop-in girders, pier shafts, and pier bases — would be built atop 2.4- to 5.5-metre-high concrete pillars. Between sets of pillars (and thus under the components) there would be a two-track concrete skidway, clad with flat steel rails upon which a Huisman would slide. Continuous belts of teflon sandwiched between sledge and rails would reduce friction to an acceptable level. Four hydraulic cylinders on the Huisman would push against four of the more than 3,500 reaction blocks built into the skidway, and force the sledge forward a little at a time.

To move a component, one of the Huismans would travel down the skidway until it was directly under the component. Then jacks on the upper part of the sledge would raise the component a few centimetres — just enough to lift it off its support pillars. The sledge would then crawl down the track at 3 metres per minute, carrying the component to the next work station, to a storage location, or to the *Svanen* for transport out into the Strait. When not carrying a load, a Huisman could lower a set of wheels and increase its speed to 60 metres per minute.

Figure 5.7 (see page 61) is a photo taken in September 1994. It shows the concrete pillars mentioned above, and some of the 6000 metres of skidway that would eventually be installed. Figure 5.8 and 5.9 shows 1996 photos of the Huisman at work moving a main girder. Since pushing a 7500 tonne load up hill made no sense, great care was taken to make the skidways in the fabrication yard level. The jetty skidway going out to the *Svanen* sloped down very slightly: 20 cm per 100 metres.

Figure 5.10 and 5.11 (see next page) show more clearly than words could, the amount of Staging Facility work that was accomplished during the summer and fall of 1994.

Work on the Approach Bridges

Because the approach bridge components were so much smaller and lighter than the main bridge components, they could be transported by truck. This made it possible to build the components for both approach bridges on one side of the Strait. The approach bridge prefabrication work was all done in a 20-acre N.B. Staging Facility in Bayfield, New Brunswick. Components needed on the P.E.I. side were loaded onto trucks in Bayfield, and the trucks were taken across the Strait on a Marine Atlantic ferry or an SCJV marine vessel.

The Bayfield facility was much simpler and less costly to create than the Borden-Carleton yard. The land used was licensed from Public Works and General Services Canada, the production area was relatively small, and there was no need to make it absolutely level. All this meant that the casting of approach bridge components could start sooner than main bridge casting. Figure 5.12 (see next page) shows how things were arranged at Bayfield.

The first pier shaft segments for the East (P.E.I.) Approach Bridge were cast in the Bayfield yard in July. While that was taking place in New Brunswick, crews on the P.E.I. side were preparing the pier sites. Early in the process, crews installed

As these before-and-after photos show, work on the P.E.I. Staging Facility progressed dramatically during the summer and fall of 1994. [Boily]

Figure 5.10 *June 3, 1994.*

Figure 5.11 *November 30, 1994.*

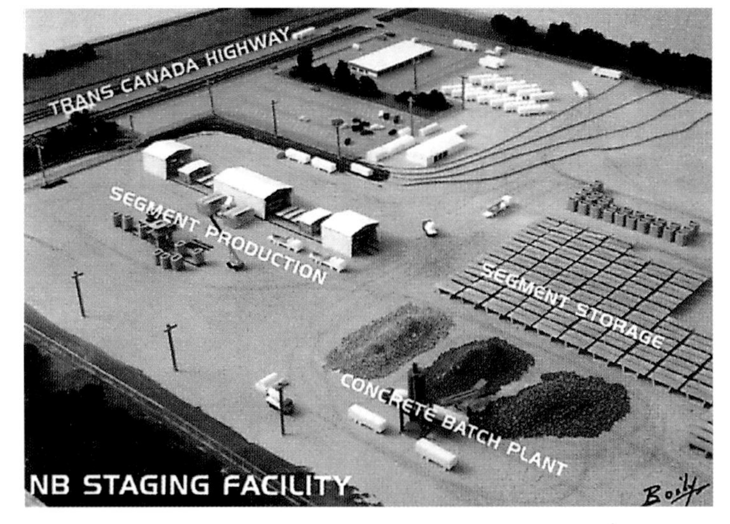

Figure 5.12 *The Bayfield facility, shown in this model, is where the approach bridge elements were cast. It was a much simpler and less costly facility than the Borden-Carleton facility. [SCDI]*

temporary cofferdams — the cylindrical sheet-steel structures shown in Figure 5.13. These cofferdams went right down into the bedrock. Then, within each cofferdam, several 2-metre diameter 5-metre deep holes were drilled into the bedrock using the drill bit shown in Figure 5.14. These became sockets for the reinforcing steel and concrete that formed the "shear keys" which secured the pier (Figures 5.4 and 5.15). Once the shear keys were in place, a conical steel ice shield was erected with a cage of reinforcing steel inside (Figure 5.16). It was then filled with concrete. (The ice shield served as a form during this process, and afterward protected the pier from ice damage.) Next, the pre-cast pier shaft ele-

re **5.13** *At some approach bridge pier sites, temporary steel dams were installed. The cofferdams went down into the k and created a dry work area within. Within this work space, proach bridge pier foundation was constructed.* [Boily]

Figure 5.14 *The first step in constructing a pier foundation for the East Approach Bridge was to drill several 2-metre diameter 5-metre deep holes using the drill bit shown in this photo. Within these holes, reinforced concrete shear keys would be cast.* [PWGSC]

Figure 5.15 *Within the cofferdam, reinforcing steel "rebar" extends upward from three just-cast shear keys.* [PWGSC]

Figure 5.16 *A crew creates the conical rebar cage that will reinforce the combined pier base / ice shield once it has been cast. The interlinked rebar ensures that the pier base / ice shield and the shear keys become one structural unit.* [PWGSC]

ments were erected atop the pier base. Finally, the internal post-tensioning tendons were tightened, pulling the pier shaft segments tightly together and forming a complete pier shaft unit. When all that had been done, the cofferdam was removed. By the end of the 1994 construction season, several fully-complete pier shafts were in place.

Design Activities

For construction to proceed, JMI/Stanley had to produce between 12,000 and 15,000 final design and shop drawings. But prior to finalizing the design details, the SCJV Design Department and JMS had to be satisfied that the bridge would meet the federal government's stringent safety requirements. Since the Borden-Carleton / Cape Tormentine ferry service would be discontinued as soon as the bridge opened, a structural failure of the bridge would isolate Prince Edward Island — an intolerable situation. In response to this concern, the government required that the Northumberland Strait bridge be significantly more reliable than North American bridges are typically required to be.

The designers faced a daunting task. They had to design a bridge that

- was many times safer than typical highway bridges in North America. (a Beta or safety index of 4.25 instead of 3.5),
- would have a useful life 2 to 3 times greater than standard bridges, and
- do it under the difficult real-world conditions of ice, wind, and marine environment that exist in the Northumberland Strait.

Over the years, standardized bridge design codes have been developed to guide engineers through the bridge design process. The engineer follows the specifics laid out in the code document and knows that a bridge so designed will meet the code's reliability or safety standard. This could not be done with the Confederation Bridge because none of the existing codes addressed this higher level of safety and longevity. Instead, the structure of this bridge, and the loads it can withstand, had to be developed from *first principles*: the laws of physics, the strength of the materials used, and the forces to which the structure would be subjected.

Engineering design involves a back and forth interplay between engineering imagination and engineering calculations. The imagination comes up with possible structures, and variations on structures, and the mathematics says whether or not they are satisfactory. In the case of a bridge, part of this process involves calculating how the bridge structure would respond to the application of certain forces or *loads* that are applied to it. Many loads and combinations of loads would have to be considered. For this particular bridge they included:

- *vehicle loads* (the weight of vehicles, and the horizontal force they exert when braking),
- *wind loads* (the force of the wind against the bridge and against vehicles on the bridge),
- *ice loads* (the force of Northumberland Strait ice against the bridge),
- *ship-collision loads* (the effect of a ship collision),
- *seismic loads* (bridge response to movement during an earthquake), and
- many others.

To calculate whether or not the design was satisfactory, 100-year-return period values for all of these loads were needed, and it was here that many of the project's 40 subconsultants played a vital role. Some of them carried out research on ice forces, wind forces, wave forces, and seismic forces. Other subconsultants investigated the geology of Northumberland Strait. The Strait seabed is bedrock covered with an overburden of till — a mixture of soil and rock left behind by the last glacier. The bedrock is not one uniform kind of rock, however. There are irregular strata of sandstone, mudstone, and siltstone of varying quality and strength. Because of this, as many as ten test holes had to be drilled at each pier location to allow geotechnical engineers to determine how far into the bedrock the dredging had to go to reach satisfactory foundation conditions. In certain instances further validation was performed using scale model tests of the foundation and pier base. Still other subconsultants did the complex work of translating the research data and the reliability requirements into specific load values that the rest of the design team could use in their calculations.

"It was the determination, changes, and concurrence of the design criteria that was perhaps the biggest challenge," explained Ross Gilmour, SCJV's Director of Design. "All the site-specific design parameters for this 100-year structure had to be derived. This was a task that under normal circumstances would have taken years to complete. But we had to complete it, subject it to the scrutiny of the Independent Engineer, and then use it to design the bridge — all in accordance with the very tight project schedule. Much of the work performed by the subconsultants amounted to scientific research; it was anything but routine engineering."

Environment-Related Activities

Strait Crossing's environmental team worked during 1993 to develop an exhaustive set of Environmental Protection Plans. These form part of its Environmental Management Plan, and detail the environmental protection procedures to be implemented by project employees during bridge construction and operation. Some involve "general" activities such as vegetation clearing, grubbing, cutting and filling for roads, blasting, infrastructure construction, solid waste disposal, concrete production, and a dozen other items. Others focus on specific Strait Crossing construction activities. There are also Environmental Contingency Plans to deal with unexpected happenings such as fuel and hazardous material spills, wildlife encounters, fires, vessel accidents, and encountering a heritage resource such as an archeological site. An environmental orientation and training program for all employees, on-site inspections, and a comprehensive reporting program all helped to make environmental concern an integral part of project life.

Figure 5.17 *In 1994, Strait Crossing installed six Osprey nesting platforms on high poles in the Cape Jourimain National Wildlife Area, through which the N.B. approach road passes. Two years later, the number of nesting Osprey pairs in the Area had more than doubled.* [PWGSC]

In addition to these protection-oriented programs and activities, Strait Crossing became involved in two wildlife habitat enhancement projects in 1993-94. The first involved the installation of six Osprey nesting platforms on high poles in the Cape Jourimain National Wildlife Area, through which the N.B. approach road passes. See Figure 5.17. In 1996, five nesting pairs were seen in the

Area, over twice the number that were there before the platforms were installed. On the P.E.I. side of the water, Strait Crossing worked with Ducks Unlimited to create a new 2.2 hectares of waterfowl habitat. "It's in a great spot," said Brian Thompson, the project's Environmental Coordinator. "There are a series of islands and channels that make it very attractive to migrating waterfowl, and Brook Trout have already migrated up from Noonan's Marsh. We named it Rollie's Marsh after a construction superintendent from Alberta who exchanged some valuable advice with me early on." See Figure 5.18

As winter settled in and 1994 drew to a close, everyone could look back at a year of accomplishment — and ahead to a lot more work.

Figure 5.18 *On the P.E.I. side, Strait Crossing worked with Ducks Unlimited to create a new 2.2 ha waterfowl habitat called Rollie's Marsh. A series of islands and channels make it very attractive to migrating waterfowl, and Brook Trout have migrated into it from Noonan's Marsh.* [Boily]

DESIGN SPECIFICATIONS

Design life:	100 years
Basic Structure:	shore to shore bridge; no causeway component
Structural Materials:	reinforced, post-tensioned concrete
Length:	12.9 km, crossing the Strait at its narrowest point
Width:	11 m from guardrail to guardrail, including one travelling lane and one emergency shoulder in each direction
Typical Elevation:	40 m above mean sea level
Navigation Span Max. Elevation:	60 m above mean sea level
Navigation Span Vessel Zone:	172 m wide and 49 m above mean sea level
Depth of Strait:	up to 35 m
Main Bridge Footings:	gravity base foundation on bedrock
Main Bridge Piers:	octagonal hollow shafts
Main Bridge Girders:	precast concrete box girders ranging from 4.5 m to 14 m deep
Main Bridge Spans:	43 spans, 10,750 m total; 250 m per typical span, plus two 165 m spans connecting main and approach bridges
P.E.I. Approach Bridge:	7 spans, 580 m in total; 93 m per typical span
N.B. Approach Bridge:	14 spans, 1300 m in total; 93 m per typical span
Approach Bridge Footings:	spread footing with drilled shear keys, or piles with pile cap
Approach Span Piers:	rectangular hollow shafts
P.E.I. Abutment Site:	just north of ferry terminal at Borden-Carleton
N.B. Abutment Site:	Jourimain Island
Utility Corridor:	a void space not less than 2.0 square metres along the entire length of the bridge which will be used to accommodate the installation and maintenance of utilities

SELLING THE FARM

If you are planning to build one of the longest bridges in the world, and

- you decide to prefabricate it on land,
- in sections weighing up to 7500 tonnes, and
- then transport those sections out on the water using a 102 metre high, twin-hulled floating crane,

you will need a rather unique place to fabricate the components. It must be spacious — not only large enough to accommodate the fabrication processes needed to make each kind of bridge component, but large enough to store several components of each kind after they are built. It must be adjacent to the water. And it must be close to where you want to erect your bridge.

There was one place that met those requirements perfectly: The 165-acre farm belonging to John L. and Anne Read — the property at Amherst Point, right next to the town of Borden-Carleton. In the real estate business they talk about the three Ls: Location, Location, Location. The Read farm had all three — and more. It was, however, a working farm. The Reads raised beef cattle, and on a summer day there would be 80 to 100 cows and nearly as many offspring roaming the 80 or so acres devoted to pasture. Would they consider selling?

It turns out that over the years many people had knocked on the farmhouse door with one proposition or another. John L. Read's grandfather had moved to P.E.I. from Pictou, Nova Scotia about 1920 and had bought the Amherst Point farm. Later, John L.'s father owned the farm. It was the father who, back in the early 1960s, received the first purchase overtures from a developer. The plan at the time was to build a causeway/tunnel/bridge across the Strait, and the Read farm was the obvious place to do some of the work. The causeway project was eventually abandoned, and the option to buy the main parcel of land was never exercised. One small parcel was sold to Maritime Electric, and another became the site of the Amherst Cove School.

"When my father owned the farm I worked on it — free labour," recalls John L. "I lived on the farm and worked at Marine Atlantic to make a living. Actually, four generations of Reads worked on the boats, from my grandfather to my daughter. I kept working there when Anne and I took over the farm in 1985. We got into raising beef cattle, and were fortunate because hay is quite accessible in this area. I'd say that farming was my main love in life."

In the 1980s a new fixed crossing project came along, and in 1989 Huang and Danczkay (the group behind P.E.I. Bridge Limited, one of the three finalists) bought an option on the farm. Things dragged on, and their option expired. Then, in December 1992, Strait Crossing came knocking, and John L. Read sold SCI a one year purchase option.

Right away, SCI began surveying the farm, and their consultant company, Jacques Whitford Environment, came to see what environmental protection measures would be needed. "As soon as we signed the option, the environmentalists came," recalls Anne Read, "the Jacques Whitford boys. Beside the farm was a marsh, and they didn't want anything from the project to go into the marsh. They were around a lot, and they were a lot of fun. These fellows would come out to do a job, but it was cold, and they'd come in to get warm. We had a big bird feeder, and here this fellow was, sitting at our table having coffee, telling us all about these birds. The SCI people were very friendly too. They

Figure 5.19 *A corner of the Read farm before it was sold. George Read, John L.'s brother, continued to live in this house after the farm was sold to Strait Crossing. He enjoyed watching activities in the P.E.I. Staging Facility from his living room window.* [George Read]

made themselves completely at home. If you made a pot of coffee they'd just get up and fill their cup. They're wonderful people. We had a good time with them."

As the option year progressed, the likelihood that SCI would be signing a deal increased — but there was never 100% assurance that it would happen. Time became an issue. SCI wanted to be moving earth the day after signing. The Reads had fields full of cows and young calves, but didn't want to sell them unless the farm actually sold.

As John L. recalls it, "We had made a deal with SCI that we were to have two or three months notice, but when the delays came because of the court cases they couldn't say yes or no. So they came and asked if we couldn't shorten this down to one month. They paid me for the inconvenience. It really helped them out, but it caused us some misery. We signed the final deal on October 7, and the month that followed was one of the most hectic months of our lives. I had to sell the cattle, I had to sell the machinery, I had to find a place to live, and vacate the property. I was worried to death about how we were going to get rid of all those cattle, but buyers came along. In truthfulness, everything fell in pretty well."

A few months later the Reads built a large house on 9 acres overlooking the West River in Dunedin, P.E.I., and now make visitors to P.E.I. feel at home in their 3-star Bed & Breakfast. "We moved into a great district," says John L., "and if you can move into a place where you're accepted into the community, it's just like staying where you were. Some people thought that my selling the farm was an awful thing to do because it had been in the family for so many generations. But when I started looking into our family history I discovered that there was never a Read generation that lived in one place for their entire lifetime. No, we were very fortunate. Life was very good to us to find such a nice place to move to."

Anne expressed similar feelings, "The farm life was a nice life. It was hard work and you didn't make that much money, but there were other pluses. I liked the animals. And there was peace and quiet too. But once the cattle left, the farm wasn't the same and I didn't mind leaving it. I was used to knowing everyone who lived nearby, and when we moved to Dunedin I didn't know anyone. That was one of the hardest things. It was also difficult to come down from a hundred and some acres to nine. But it's nice down here. It's quiet. And it turns out that our neighbor at the top of the hill is from Borden. And the lady down at the corner — her mother taught John L. in school in Borden. And the guy next door worked with him on the boat. So you soon settle in."

Figure 5.20 *Livestock and the* Abegweit *the summer before the farm was sold.* [John L. Read]

CHAPTER SIX

A YEAR OF MILESTONES — 1995

The year 1995 was a year of validation and confirmation — a year of proof that the Confederation Bridge could be built, and built in a most innovative, sometimes spectacular way. During the year every kind of component was successfully fabricated, and at least one of each kind was successfully placed in the Strait. It was therefore a year punctuated by many firsts. This chapter reviews, step by step, the way the bridge was built. It reviews the components that were fabricated and the marine procedures that were followed — and touches on some of those milestones.

Before the first construction milestone, however, Strait Crossing received a major pat on the back for the way it was handling the environmental aspect of the project. On March 9, Strait Crossing and Jacques Whitford Environment Ltd. were awarded the prestigious 1994 *Environmental Achievement Award* by the Canadian Construction Association for "The far-reaching and exhaustive environmental consideration embodied in the bridge project between Prince Edward Island and New Brunswick." The CCA's annual award recognizes innovative or exemplary

Figure 6.1 *Workers construct a rebar cage for one of the main bridge pier bases.* [PWGSC]

Figure 6.2 *Main bridge concrete was prepared in this concrete plant at the Amherst Point end of the Borden-Carleton yard.* [Boily]

environmental conduct, and the Strait Crossing project was the unanimous choice of the Association's Environment Committee for the 1994 award.

Main Bridge Component Fabrication

The P.E.I. Staging Facility was completed in April, and by then many fabrication-related activities were already under way. These included the readying of forms for concrete pouring, and the construction of "rebar" cages — lattice-work structures of steel reinforcing bars that are embedded within cast concrete components to strengthen them. Figure 6.1 (see previous page) shows workers constructing a rebar cage for one of the pier bases.

High Performance Concrete

Concrete is a mixture of binding paste (normally Portland cement, water, and air) with aggregates (normally sand, and gravel or crushed stone). There are several varieties of Portland Cement, many sizes and types of aggregate, and many other possible additives such as *fly ash* (a byproduct of coal combustion), *silica fume* (a byproduct of silicon-boron alloy production), and *superplasticizers* (used to improve workability).

High Performance Concrete refers to concrete that has one or more especially desirable characteristics such as higher than ordinary strength, high durability, high resistance to abrasion, or low absorption of chemicals from the environment. Exceptional durability was especially important on this project. During the next 100 years, some parts of the bridge will be subjected to multiple freeze/thaw cycles, others will be immersed in salt water or covered with salt spray, and still others will be subject to ice abrasion. The main cause of concrete structure deterioration is corrosion of the reinforcing steel, so not only must the concrete itself be durable, but it must protect the rebar within. Silica fume helps provide this protection by raising the electrical resistivity of the concrete and by inhibiting chloride ion migration from the salt water on the outside to the rebar on the inside. All the Portland cement used on this project contained at least 7.5% silica fume.

A total of 478,000 cubic metres of concrete was placed during the course of the project. The bridge concrete fell into three classes. All three had high durability, but they differed in other characteristics. The Class A structural concrete used for main bridge components had especially high strength — 55 MPa after 28 days of curing compared with 20 to 35 MPa for general-use concrete. The Class C concrete, which used 3 parts fly ash to each 7 parts of cement, had a lower than usual temperature rise during cure. It was used in large solid components such as the approach pier foundations. The Class F or "Tremie" concrete was specially formulated for underwater placement. All these mixtures were prepared in the concrete plant at the Amherst Point end of the Borden-Carleton Yard — Figure 6.2. Concrete was also prepared at the Bayfield site.

Matchcast Templates

The matchcast template — often called the capital-M *Matchcast* for short — is a relatively uncomplicated main bridge component, and was the first main bridge component to be cast (July 1994). But this 5 by 10 by 1 metre rectangle of concrete played a critical role in bridge assembly. Figure 6.3 shows both the Matchcast and the component it mates with — the central *hammerhead* section of the main girder. The hammerhead's role in the overall scheme is shown in Figure 6.4 (see next page).

Once a Matchcast was cast and cured, it was temporarily integrated into the casting formwork for the hammerhead. Thus, when the hammerhead was cast, the underside of the hammerhead acquired the shape of that particular template. Later, the hammerhead would be extended into a full-sized main girder, and a pier base and pier shaft would be installed in the Strait. At that point the matchcast template would be affixed to the top of the pier shaft with exactly the right

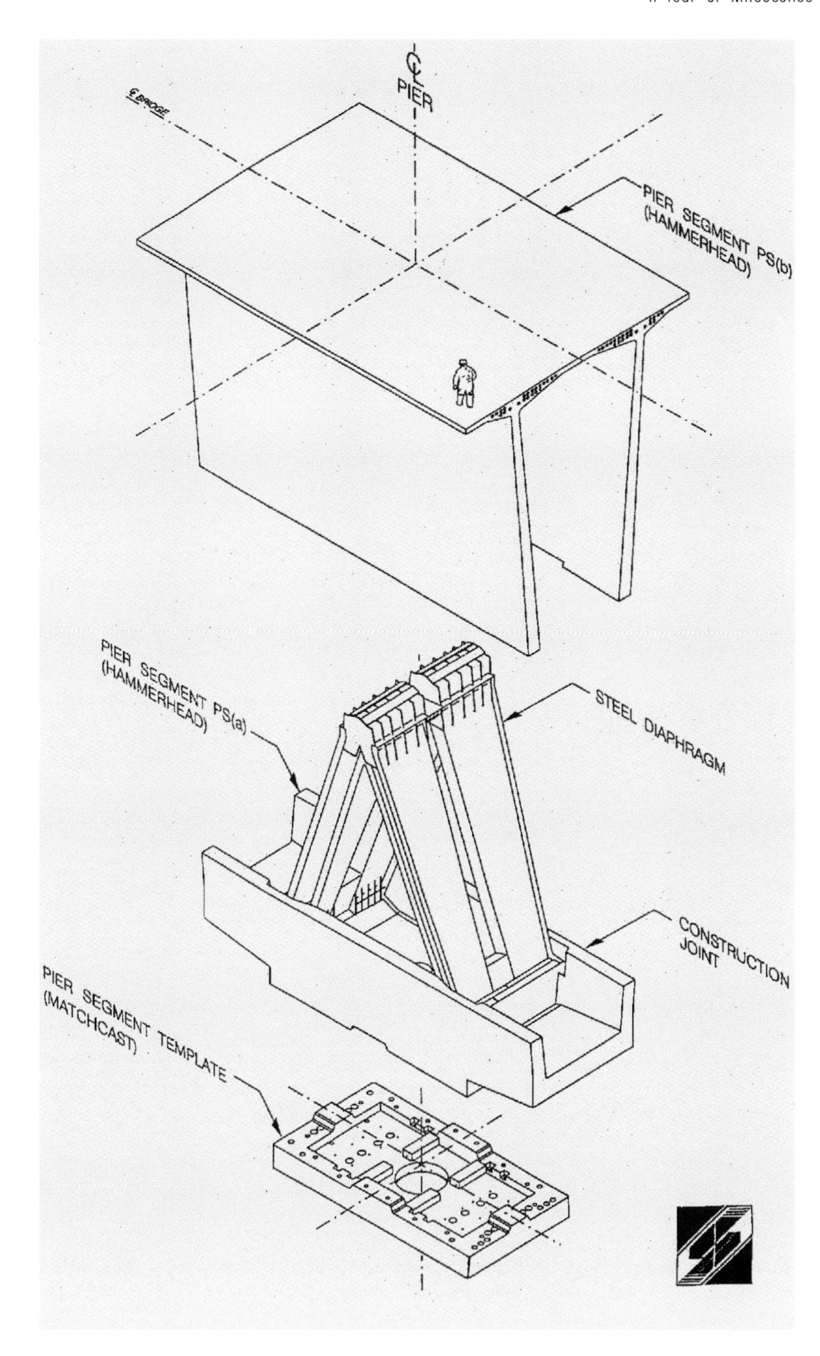

Figure 6.3 *Shown here are both the Matchcast Template and the component it mates with: the central* hammerhead *section of the main girder.* [SCDI]

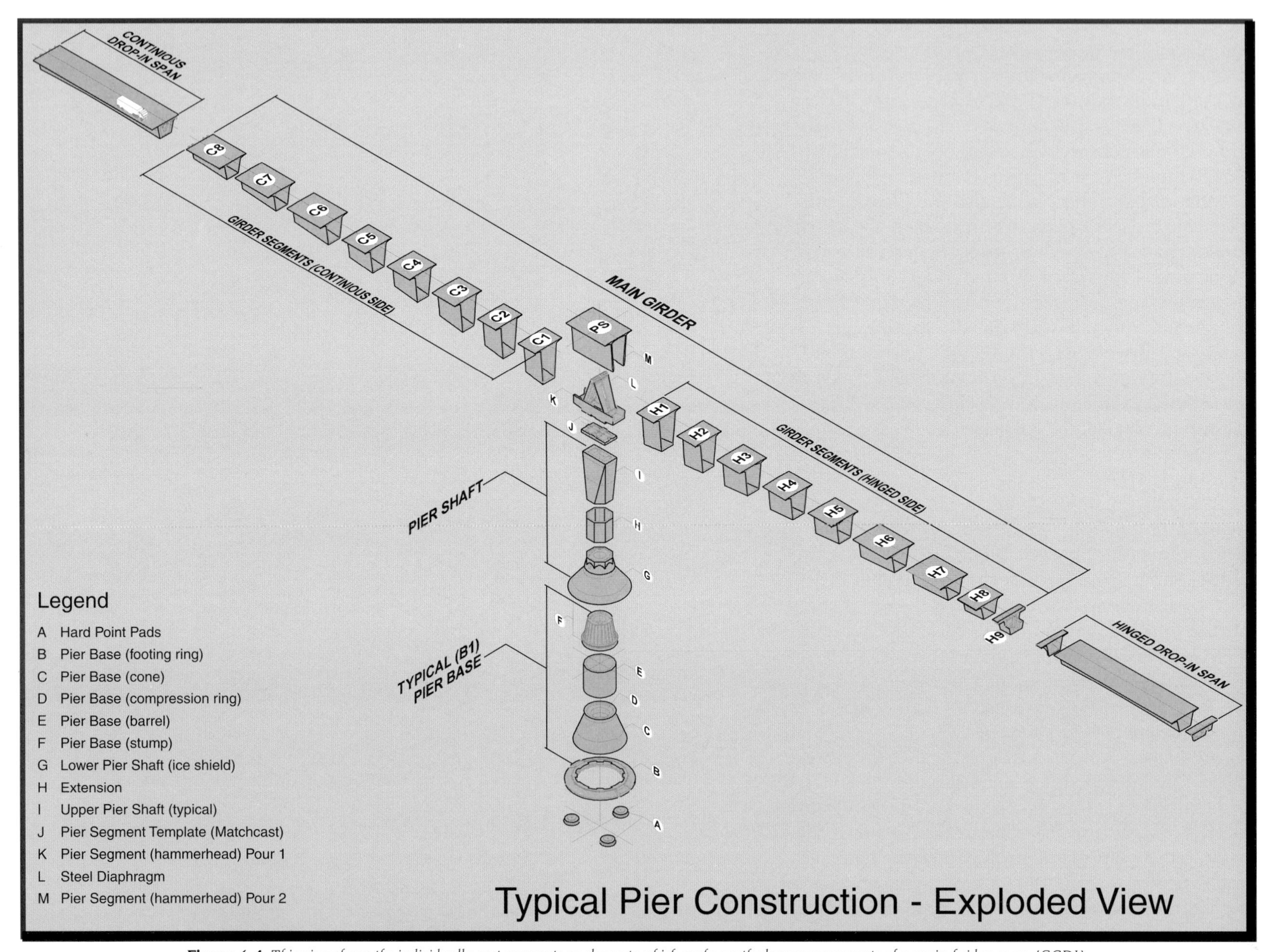

Figure 6.4 *This view shows the individually cast segments or elements which make up the larger components of a main bridge span.* [SCDI]

Figure 6.5 *Several of the 4.5 m diameter 0.6 m high* hard points *used to support main bridge pier bases until a complete pier-base-to-bedrock connection could be made with tremie concrete.* [PWGSC]

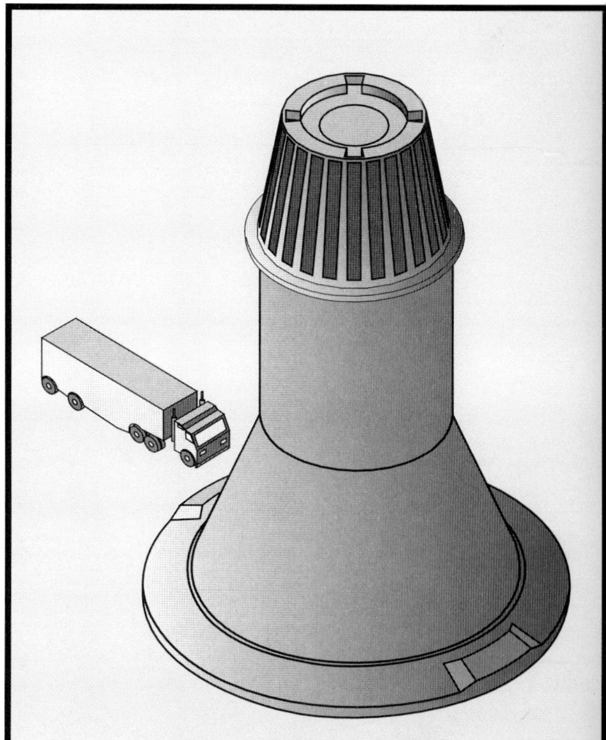

Typical (B1 type) Pier Base

Figure 6.6 *A complete main bridge pier base. It was progressively cast, segment by segment, starting with the footing ring and working upward.* (*See also Figures 6.4 and 6.7*) [SCDI]

orientation. The main girder would then be lowered onto the shaft. The bottom of the girder and the top of the Matchcast would nest into each other perfectly, and because of that, the whole girder would be properly oriented. By carefully aligning and fixing a 100 tonne part, Strait Crossing had avoided the much more difficult job of aligning and fixing a 7500 tonne part.

Hard Points

The hard points are other small, simple components that do a vital job. Shown in Figure 6.5, they are 4.5 metres in diameter, about 0.6 metres high, and are dished-in on the underside. Three of them, accurately placed on dredged bedrock at a pier site, provide temporary support for the pier base until the space between bedrock and the underside of the base can be filled with tremie concrete.

Pier Base — First One Completed on May 4

Figure 6.6 is a drawing of a completed pier base. Figure 6.4 reveals that this component was not cast all at one time, but in four separate casting opera-

Figure 6.7 *Sectionalized forms, such as this one, allowed large components to be cast from the bottom up, segment by segment.* [Boily]

Figure 6.8 *Shown here are various stages of production in the pier base assembly area and the main girder assembly area.* [PWGSC]

tions, section by section. First the bottom ring is cast, then the cone, next the cylinder, and finally the top or stump segment. Sectionalized forms, such as the one shown in Figure 6.7, allowed the component to be cast from the bottom ring up in this segment-by-segment way. Figure 6.8 shows three production lines devoted to pier base fabrication, and pier bases at many different stages in the casting process.

Pier Shaft (with Ice Shield) — First One Completed on June 3

The complete ice shield and pier shaft assembly appears in Figure 6.9. If we refer again to Figure 6.4, we can see that two casting operations were involved. The lower ice shield portion was cast first, then the pier shaft.

The ice shield is a conical structure 20 metres in diameter at its base. The cone is 13 metres high, and extends roughly 4 metres below the water and 9 metres above. Its purpose is to reduce the amount of force that moving ice would otherwise exert against the bridge piers. The force is reduced because the impacting ice tends to ride up the 52 degree slope of the cone and break under its own weight. Ice that is flat and relatively thin, Figure 6.10, (see next page) breaks up smoothly and reliably. But even with quite heavy ice conditions, like the ridged ice of Figure 6.11 (see next page), an impact against the sloping ice shield is preferable to an impact against a vertical pier.

The ice shields for piers P1 to P11 were clad with a 1 cm thick steel skin so that the concrete would not be abraded by the ice. Later ice shields were made of ultra high strength concrete, and the steel skin was no longer used.

Drop-in Girder — First One Completed on June 24

The diagram of Figure 6.12 (see next page) shows the drop-in girders, and how they connect one main girder to the next. There are two types of drop-in spans, *continuous* and *hinged*, and these two types of connection alternate for the length of the bridge. The continuous drop-in girders are firmly attached to the main girders on either side — making, in effect, a one-piece structure. The hinged drop-in girder, on the other hand, rests on fixed bearings at one end and sliding bearings at the other. The sliding bearings allow the bridge to expand and contract freely in response to temperature changes, or changes in the concrete itself over time. This sliding connection at every second span

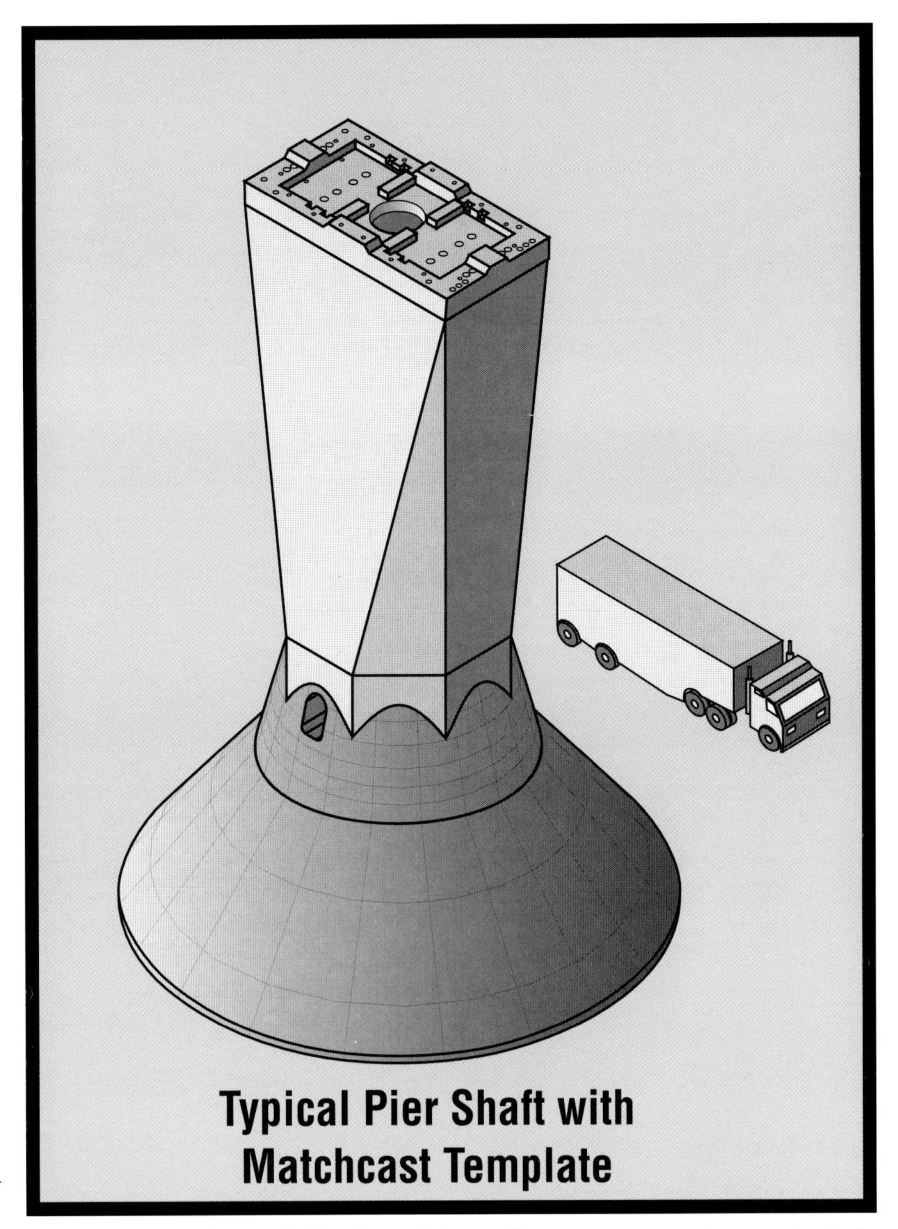

Figure 6.9 *A complete ice shield and pier shaft assembly. Two casting operations were involved. The lower ice shield portion was cast first, then the pier shaft.* [SCDI]

Figure 6.10 *Strait ice is shown here moving past the bridge. As the ice sheet is forced up the 52 degree slope of the ice shield cone, it breaks under its own weight and the broken pieces pass on both sides of the pier.* [Boily]

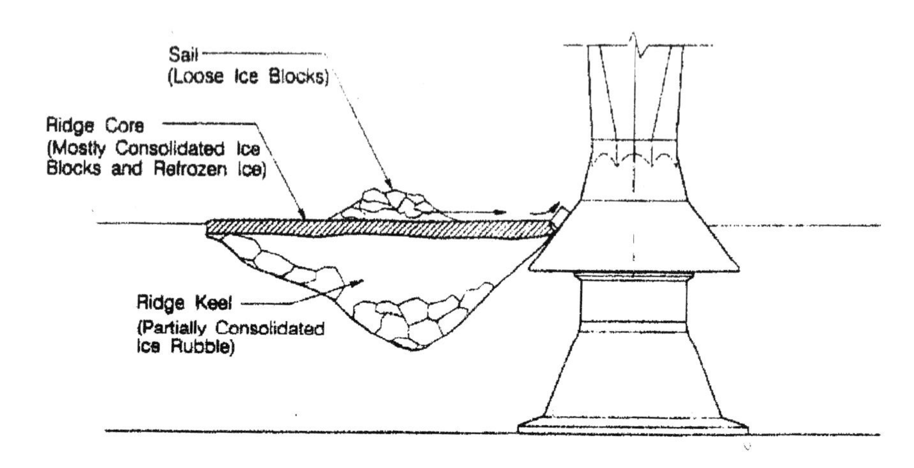

Figure 6.11 *Strait Ice is sometimes heavy and ridged, but even when it is, an impact against the sloping ice shield is preferable to an impact against a vertical pier.* [JMI]

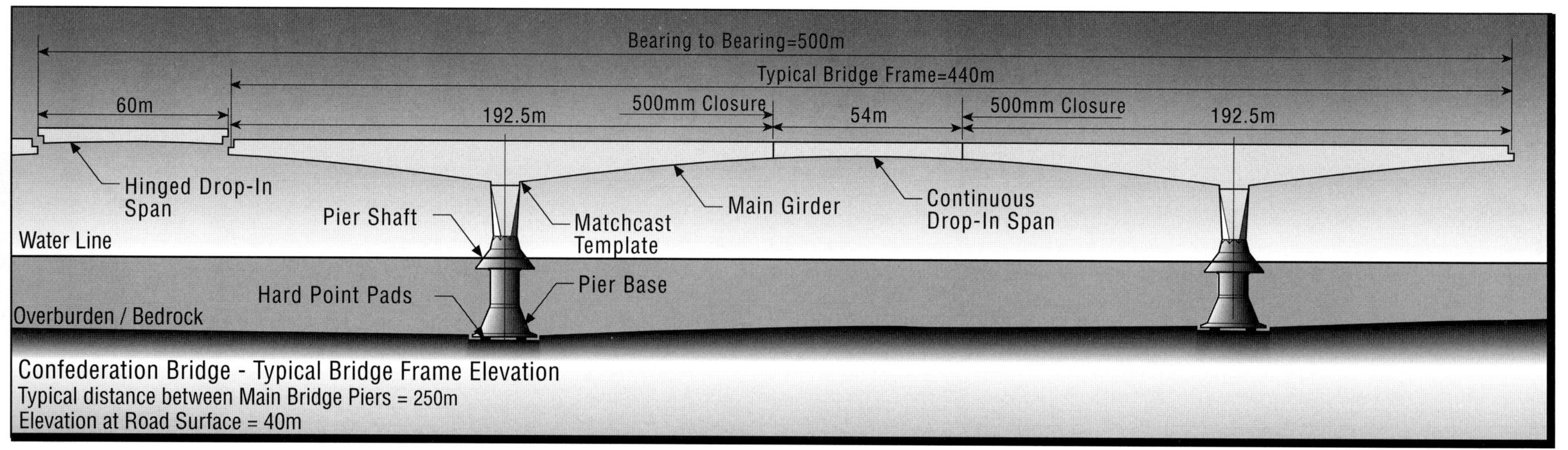

Figure 6.12 A *bridge* frame *involves two piers, their main girders, and the continuous drop-in girder that connects them.* The continuous drop-in girders *are firmly attached to the main girders on either side — making, in effect, a one-piece structure.* The hinged drop-in girder, *on the other hand, rests on fixed bearings at one end and sliding bearings at the other. The sliding bearings allow the bridge to expand and contract freely in response to temperature changes or changes in the concrete itself over time.* [SCDI]

also ensures that a catastrophic failure of one span will not cause a domino effect and bring others down.

Main Girder — First One Completed on August 9
With a length of 192 metres and a weight of 7500 tonnes, the main girder is the largest and heaviest of the components. It is also the most involved piece to cast. Constructing a main girder starts with constructing the hammerhead section shown in Figures 6.3 and 6.13 (see next page). The hammerhead contains a large steel structure in the shape of an inverted V known as the diaphragm. It makes this critical section of the structure strong and stiff, and is the anchoring point for many of the post-tensioning tendons that tie the main girder to the pier shaft and drop-in girders.

Figure 6.14 (see next page) shows what happens from that point on. Once the hammerhead has been cast, a Huisman moves it to the next position in the production line, and a new girder segment is cast at each end of the hammerhead. That assembly is then moved to the next position, and two more segments are added. This happens again and again until eight segments have been added to each side of the hammerhead. Next, the girder is moved into the girder storage area where a special hinge segment is added to one end — the end that connects to the hinged drop-in girder. Completed girders remain in the storage area until the *Svanen* is ready to place them.

Main Bridge Marine Work

The major marine work in 1995 was done by three vessels: the *Betty-L*, *Buzzard*, and *Svanen*.

- ***Betty-L*** A large (128 m x 30 m x 7.6 m hull) vessel used for mass excavation, final dredging and cleanup, and placing hard points at pier locations using a special hard point frame suspended from the vessel.
- ***Buzzard*** A 43 m x 30 m jackup barge used primarily for offshore concreting operations.
- ***Svanen*** The 103 m x 72 m x 102-m-high heavy lift vessel used to transport and place main bridge components. Design load capacity: 8200 tonnes.

Knowing Where to Put It: GPS
Most of us think of surveying as something done to establish the boundaries of a piece of property, but in construction work surveying is much more than that. In that context, surveying means accurate positioning in the most general sense. It means making sure that constructions of all kinds are located where they should be — vertically as well as horizontally. Traditional "autotracking" optical surveying techniques could have been used on this project if the bridge had been built progressively out from the shore, and if the surveying didn't have to be done at night or under foggy conditions. These were unacceptable constraints, so an alternative was needed.

The most promising alternative was the Global Positioning System, or GPS, that uses radio signals from several satellites to determine position and altitude. Strait Crossing's Mirek Bursa was responsible for the design and implementation of the surveying aspects of the project, and recalls what happened: "In 1992, when we had to decide what system to use, GPS technology wasn't accurate enough. The project design required a placement accuracy of 2 cm, and the best GPS could do at the time was 30 to 50 cm. So we started talking to the GPS gurus who were working on improved versions of the system. The accuracy was not yet there, but it looked like it eventually would be. We felt we had to

Figure 6.13 *Constructing a main girder starts with constructing the hammerhead section shown above. The hammerhead contains a large steel structure in the shape of an inverted V known as the diaphragm. It makes this critical section of the structure strong and stiff, and is the anchoring point for many of the post-tensioning tendons that tie the main girder to the pier shaft and drop-in girders.* [PWGSC]

Figure 6.14 *Once the hammerhead has been cast, a Huisman sledge moves it to the next pos tion in the production line, and a new girder segment is cast at each end of the hammerhead. Th assembly is then moved to the next position, and two more segments are added. This happe again and again until eight segments have been added to each side of the hammerhead. Final the girder is moved into the girder storage area where a special hinge segment is added to one e — the end that connects to the hinged drop-in girder.* [Boily]

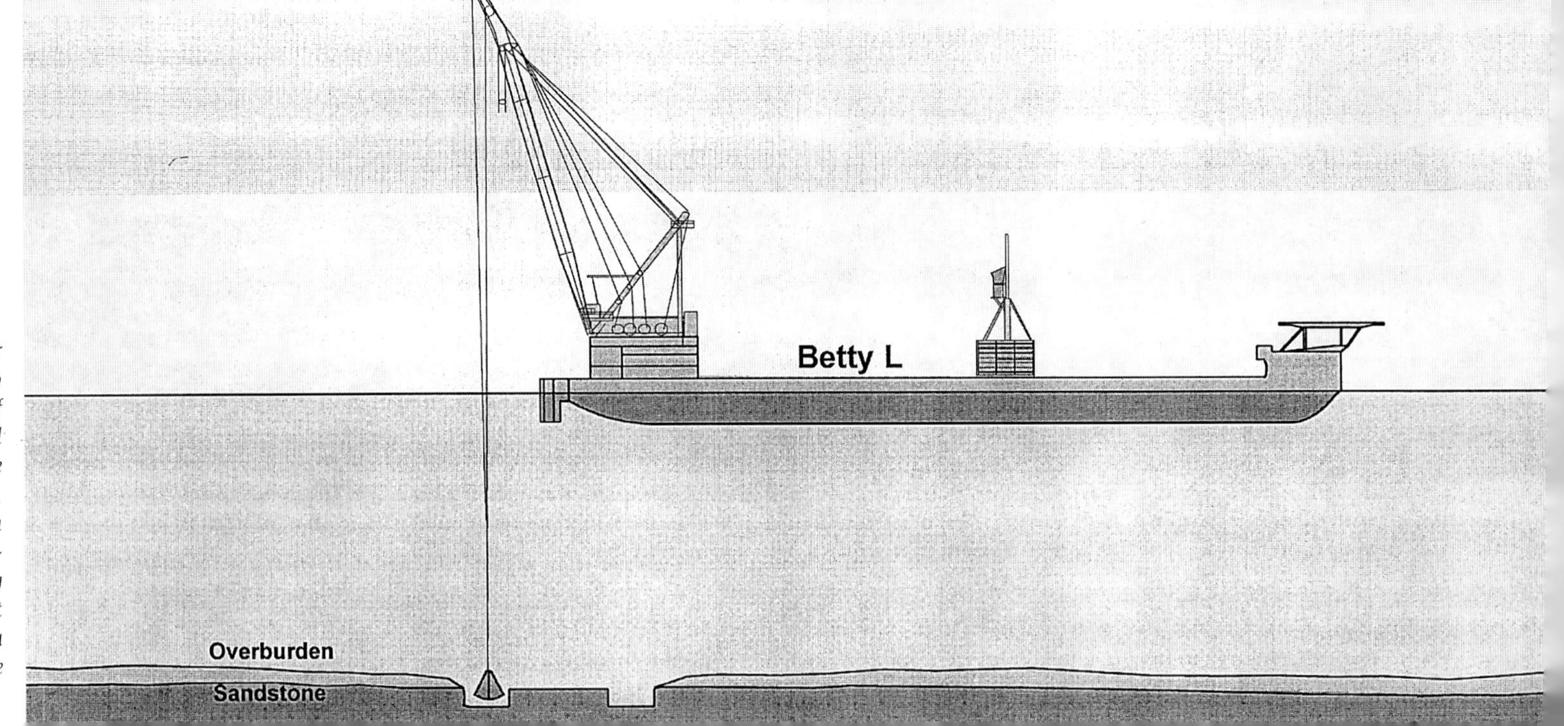

Figure 6.15 *Dredging at pier base locations took place in two steps. First, the overburden of glacial till was removed — the soil and rock which covered the bedrock at the pier location. Second, a template was placed on the seabed at the exact pier location, and a crane on the dredging barge moved a clamshell bucket around the template, dredging a circular, 6-m-wide trench in the bedrock.* [SCDI]

Figure 6.17 *Divers were used to check the underwater work.* [PWGSC]

Figure 6.16 *The frame held a hard point at the bottom of each leg. Under the hard points were bags connected to hoses. A crane lowered the frame into the water, and GPS positioned it so that the three hard points were within the circular trench and at the desired height above the bottom of the trench. A concrete mixture called grout was then sent down the hoses into the bags, filling them so they made contact with both bedrock and the underside of the hard points.* [PWGSC] trust the future. We had to trust that developments would take place and the accuracy we needed would be there when we started placing components in the Strait." That trust was rewarded. By the summer of 1995 Strait Crossing had state-of-the-art "real time kinematic mode" GPS equipment that allowed components to be placed anywhere in the Strait with 2 cm accuracy.

Dredging Pier Base Foundations

The bottom of each B1 pier base is a flat ring-like surface having an outside diameter of 22 m, an inside diameter of 14 m, and a width of 4 m. In placing the pier base on the seabed the objective was to create a stable, uniform, intimate connection between that surface and the bedrock under it. That was accomplished through a three-step process of dredging, hard point placement, and tremie concrete placement.

To minimize environmental impact and cost, the dredging on this project was both precise and minimal. It took place in two steps. Step one was to remove the glacial till —the soil and rock which covered the bedrock at the pier location. Step two was to create a circular trench in the bedrock, 6 m wide, with inner and outer diameters of 12 m and 24 m. To do this, a template was placed on the seabed at the exact pier location, and a crane on the dredging barge moved a clamshell bucket around the template, dredging a circular trench in the bedrock to the depth determined by earlier test drilling. See Figure 6.15 (see page 82). Both sonar imaging and divers were used to monitor what was happening below.

The dredged material was put into scows and transported to a designated site off Amherst Cove — one that lobster fishers had not previously found productive. In an attempt to enhance the site's attractiveness to lobster and rock crab, the dredged material was deposited in a controlled way. Loads of fine material were capped with loads of boulders and rock, and the large material was evenly deposited. This reduced the potential for fine material to drift away, and created a submerged artificial island having the sort of boulder-and-crevice seabottom environment that lobsters and rock crab seem to like.

Placing Hard Points — First Ones Placed on June 2

As Figure 6.16 (see previous page) indicates, the *Betty-L* placed hard points with a special 3-legged frame. At the bottom of each leg, the frame held a hard point. And under the hard points were bags connected to hoses. A crane lowered the frame into the water, and GPS positioned it so that the three hard points were within the circular trench and at the desired height above the bottom of the trench. At this point a concrete mixture called *grout* was sent down a hose into one of the bags, filling it so that it made contact with both bedrock and the underside of the hard point. The other two bags were then filled with grout, and the frame was checked to see that all three hard points were level with each other. The frame remained attached to the hard points until the grout hardened sufficiently. After the frame was removed, a diver gave the hard points and grout bags a final inspection and removed the hoses.

Placing Pier Bases — First One Placed on August 7

Anchored by six sea anchors and two lines to the jetty, the *Svanen* picked up a pier base from atop a Huisman. See Figure 6.18. The *Svanen* then travelled to the pier base location (guided in final positioning by GPS) and was anchored firmly into that position with 8 pre-set sea anchors. A

Figure 6.18 *The* Svanen *prepares to transport and place the very first pier base.* [PWGSC]

diver inspected the trench containing the hard points, and if it needed to be cleaned out, that was done with equipment on the *Svanen*. The pier base was then lowered onto the hard points, and the *Svanen* left. Next, the *Buzzard* arrived and divers descended. They pumped tremie concrete into the trench until it filled the space between the bedrock and the undersurface of the pier base.

Figure 6.19 *The* Svanen *places the first main girder on* October 1, 1995. [PWGSC]

Placing Pier Shafts — First One Placed on August 23
After picking up a pier shaft and transporting it to the pier location, the *Svanen* carefully lowered it over the top end of the pier base. The pier shaft was not set directly on the pier base, but the two were held slightly apart by jacks. The narrow space between the components was later filled with grout, creating a solid, load-bearing connection.

Placing Matchcast Templates — First One Placed on August 23
The matchcast template was normally placed on the pier shaft immediately after the shaft had been placed. The *Svanen* would back away from the pier location so that the barge carrying the Matchcast could position itself between the *Svanen*'s two hulls. The *Svanen* then lifted the Matchcast from the barge, raised it to a level higher than the top of the pier shaft, and returned to the pier shaft location. The template was then lowered onto 4 pre-installed jacks on top of the pier shaft. The jacking system would be very carefully adjusted until the matchcast template was in the correct position. Workers packed small concrete blocks into the joint between the shaft and the Matchcast, and then pumped grout into the remaining space.

Placing Main Girders — First One Placed on October 1
Picking up and moving a main girder with the *Svanen* was similar to moving other components, but with some additional checks needed during the lifting process. Lowering the girder onto the matchcast template also required exceptional care. When the main girder had been lowered to within a metre of the Matchcast, the crew applied epoxy to the top surface of the template. Then, as the girder was lowered that last metre, the crew carefully monitored its position to ensure that the two mated perfectly. Figure 6.19 shows the *Svanen* placing the first main girder.

Figure 6.20 *The approach bridges were built from the shore out, and the equipment which made that possible was the 144-metre-long Italian-made launching truss shown here. This device balanced on the pier table with its back end on the already-built bridge deck and its seaward end extending out beyond the end of the deck.* [PWGSC]

Placing Drop-in Girders — First One Placed on November 17
Here, too, the basic procedure was similar. Final positioning was the difficult part, and it was somewhat different for the two types of drop-in girder.

Figure 6.21 *A gantry winch atop the launching truss lifted bridge deck segments from the trucks that delivered them, and moved these components to the end of the approach bridge.* |*Boily*|

Post-tensioning

Most of the operations just discussed were followed by some sort of post-tensioning activity within the hollow components. Just as post-tensioning tendons were used to tie separate sections of a girder into a single functional unit, they were also used to tie girders to pier shafts, pier shafts to pier bases, and continuous drop-in girders to the main girders on either side. In other words, post-tensioning was used to add strength to the entire structure, and make each main bridge span a continuous unit. It took 7 high-tensile-strength steel wires to make a 15 mm diameter post-tensioning *strand*, and from 4 to 31 post-tensioning strands to make a post-tensioning *tendon*. Jacks were used to tighten the tendons, and both jack force and tendon elongation were measured to ensure that each tendon was being stressed to the proper degree.

Approach Bridge Activities

Approach bridge casting continued during the year in the Bayfield Yard, and on the P.E.I. side, approach bridge erection began as soon as weather permitted. The approach bridges were built from the shore out, and the tool which made that possible was the 144-metre-long Italian-made launching truss shown in Figure 6.20 (see previous page). This device balanced on the pier table with its back end on the already-built bridge deck and its seaward end extending out beyond the end of the deck. A truck carrying the next deck segment would back in close to the landward end of the truss. A gantry winch designed to travel along the top of the truss then lifted the piece off the truck — as shown in Figure 6.21. Meanwhile, workers coated the exposed end of the previously installed segment with epoxy. The truck drove away, and the gantry moved down the truss toward the sea, carrying the deck segment with it. When the segment had been carried past the end of the bridge, it was lowered and turned 90 degrees to bring its matchcast face against the epoxy-covered end of the bridge. (See Figure 6.22) Temporary post-tensioning bars were tightened, firmly joining the two sections. Then permanent post-tensioning tendons were installed. The approach bridge was now a little bit longer, and ready to receive the next deck segment.

Figure 6.22 *When the gantry crane had carried the bridge deck segment past the end of the bridge, the segment was lowered and turned 90 degrees to bring its matchcast face against the epoxy-coated bridge-end surface. Temporary post-tensioning bars were then tightened, firmly joining the two sections.* [PWGSC]

Figure 6.23 *To build the approach bridge on the New Brunswick side, a temporary work surface was installed which allowed trucks and cranes to drive to several pier sites.* [PWGSC]

In New Brunswick, work was underway to install the piers for the West Approach Bridge. Because of the very shallow water and the experience gained constructing the East Approach foundations, cofferdams weren't used for the nine piers closest to shore. Instead, a temporary work surface was installed which allowed trucks and cranes to drive to those pier sites — see Figure 6.23. The pier foundations also differed, as Figure 6.24 indicates. Here, 6- to 11-metre-long, 2-metre-diameter reinforced concrete piles would be used instead of the shorter shear keys.

There were several approach bridge firsts during the year. They included:

- **June 2 — First P.E.I. Approach Span Completed**
- **August 7 — First N.B. Pier Completed**
- **October 11 — P.E.I. Approach Bridge Completed**

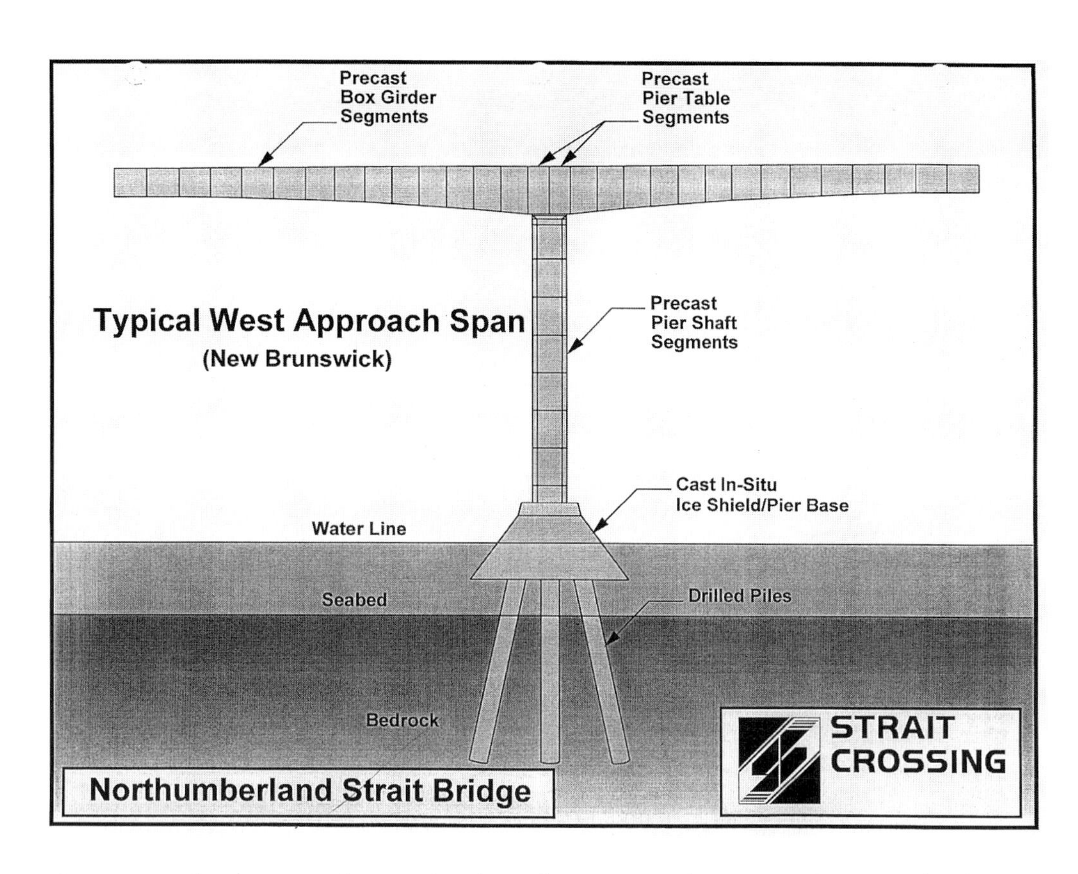

Figure 6.24 *The New Brunswick Approach Bridge used 6- to 11-metre-long, 2-metre-diameter reinforced concrete piles instead of the shorter shear keys used in the foundations for the P.E.I. Approach Bridge.* [SCDI]

Figure 6.26 *A worker fabricating rebar.* [Boily]

Figure 6.25 *Workers put the finishing touches in the rebar for the upper portion of the pier shaft.* [PWGSC]

Figure 6.28 *Workers are dwarfed by massive bridge girders.* [PWGSC]

Figure 6.27 *Once the rebar has been bent and tied, and the form is in place, the concrete is poured.* [Boily]

Demands of the Schedule

During 1995 every element of the construction system was tested, and every one worked. And by the end of November there were pieces of every kind out in the Strait. But there weren't a large number of pieces out there. The schedule loomed. To be sure of meeting the June 1, 1997 date for bridge opening, every remaining bridge component would have to be fabricated and placed by the end of 1996. Was that possible?

Figure 6.29 By October 1995 *progress had been made, but there was still much to be done if the bridge was to be finished on time.* [Boily]

THE SVANEN RETROFIT

In 1993 Strait Crossing purchased the heavy lift vessel *HLV Svanen* from the European Storebælt Group in Denmark where it had just helped complete the Storebælt project's 6600 metre West Bridge. Built in 1990, the *Svanen* had lifted and placed 108 metre long, 5000 tonne girders on that project.

To meet the needs of the Northumberland Strait Bridge project, the *Svanen* had to be modified. The vessel's height had to be increased from 76 metres to 102 metres, width increased from 65 to 71.8 metres, lifting capacity raised from 7000 to 8500 tonnes, and propulsion system improved.

Strait Crossing wanted this work to be done in Atlantic Canada, and approached several Canadian shipyards — but no Canadian firms bid on the job. One of the problems was the lack of drydocking facilities that could handle the *Svanen*'s width. The winning bidder was a French firm, Montalev, and they came up with a novel solution to the problem. Wout Krispijn of Ballast Nedam was there: "In October, 1994 the *Svanen* was towed to Dunkirk in the western part of France. There was no drydock wide enough, but Montalev came up with the beautiful solution of using the tides in the harbour. They prepared a flat surface on the beach, and on December 2, 1994 the *Svanen* was floated in on what is called a "spring tide" — a tide that swings between exceptionally high and exceptionally low, and occurs around the time of the full moon and the new moon. Twenty-four bulldozers made a dam, the water flowed out at low tide, they closed the dam, and there it was, sitting on the beach."

A crew of 250 people worked continuously through the winter and spring to complete the work. In March, the top of the gantry — weighing 1100 tonnes — was raised by 26 metres. "In the beginning of May," recalls Wout Krispijn, "there was another spring tide coming, and we were far enough along that we could refloat the *Svanen*. The bulldozers came back, the spring tide came in, and it floated. There were no leaks, and using the *Svanen*'s propulsion apparatus we travelled to the other side of the harbour. There we finished all kinds of details. We had to do a lot of functional testing, all of which was successful.

"In the middle of June we were able to load it onto a wide transport pontoon that we found in Norway, and at the end of the day we had 11,000 tonnes of steel out of the water. The *Svanen* was welded tight to the transport pontoon, and in the first week of July two tugboats of 15,000 horsepower each hooked up and they left on the trip to Canada. We estimated an average speed of 5 knots, and a travel time of 35 days, but it went a lot quicker than we calculated due to beautiful weather. The whole North Atlantic was a lake rather than an ocean. So they sometimes reached a speed of 10 knots, and they arrived here 17 days after departure."

The *Svanen*'s arrival in P.E.I. waters on July 13 was greeted with great excitement and much media coverage. This was a major event, and Islanders aplenty headed for the shore to watch the *Svanen* pass down Northumberland Strait toward Borden-Carleton. The giant vessel was separated from its transport barge on July 21, went through a series of lifting tests, and on August 7 successfully placed the first main bridge component — a pier base.

Figure 6.30 *The* HLV Svanen, *was pulled by two tugs across the Atlantic. There was great excitement when it arrived in* P.E.I. *waters.* [PWGSC]

TOWARD COMPLETION — 1996

As 1995 drew to a close, project management personnel were both pleased and concerned. One of each component had been fabricated successfully, and every construction and erection method had been proven. Yet for the final phases of certain activities the "cycle time" — the time required to carry out the activity — had yet to be optimized.

To meet the bridge-completion schedule this optimization was necessary, and meeting the schedule was a serious matter. If the bridge did not open on schedule June 1, 1997, the consortium would have to subsidize ferry operations from that date until the bridge did open — an expensive prospect. Were there ways of addressing the troublesome situations? Ways of reducing the cycle times?

Some Changes

There were ways, and they involved the addition of resources. In late 1995, the position of Construction Manager was filled by the highly experienced Gilles deMaublanc from GTM. In early 1996, Harmen Blom came from Ballast Nedam to fill the newly established position of Project Manager. "Paul Giannelia, Harmen Blom, and I became a small management team, and it worked very well," recalled Gilles deMaublanc. With the departure of Morrison-Knudsen, the position of Marine Works Manager also opened, and was filled by Rene Kolkman from Ballast Nedam.

"Gilles very quickly identified some of the critical areas," recalls Ross Gilmour, Strait Crossing's Director of Design. "The dredging was one of these. If erection was speeded up as much as it needed to be, dredging would have to be speeded up too. Another concern was the closure joint on the continuous drop-in girder. These joints were originally 1.5 metres long, and the first one had taken about four weeks to do. We went down to a half metre joint, and ultimately they were pouring those joints within 24 hours of setting the girder. We'd also been working to revise the ice shields. The ice shield started with a steel skin, but using steel in this application was very time consuming and expensive. So we spent the winter of 1995-96 developing a different material to use for the ice shields. We ended up developing a very high strength, high durability concrete mix. It was unprecedented in the industry in

Figure 7.1 *The winter construction effort was a great success, as this April 14, 1996 photo of the P.E.I. Staging Facility shows. Marine work started shortly after, with the first component of the 1996 construction season being placed on April 20.* [PWGSC]

terms of strength, durability, and freeze/thaw resistance for concrete produced in a high-production environment. There were also major changes needed in rebar fabrication to make that part of the job as efficient as possible, and changes were needed in post-tensioning."

A Successful Winter

These design changes would help speed things up, but it was also clear that component fabrication had to continue throughout the winter. This would be difficult because casting concrete under winter conditions requires special measures. Harmen Blom recalled the situation: "To boost production in the yard we had to construct housing around the elements, provide heating, and so forth. In addition, the people on the job had to put up with difficult working conditions. I was very impressed with the way the Canadian workers were able to work in the very cold conditions during the winter, and get the job done."

The winter construction effort was a great success. Cycle times shrank and became more nearly equal as each crew gained experience and progressed along its learning curve. By mid-April the fabrication yard was full. (See Figure 7.1) Ross Gilmour described the success this way: "This kind of an operation is much like a factory. It's building widgets. If you can design all the widgets the same, and give the guys the tools and the instructions on how to build them, it's amazing how good they can get at it. If you do the key things up front: detailing the concrete so the forms can be stripped, detailing the rebar so it can be preassembled in place, detailing the post-tensioning so it can be stressed sequentially, and so on, the next thing you know, it starts to click. And that's what happened in early 1996. We had tweaked every piece of it all the way along,

Figure 7.2 *Some work in the fabrication yard continued around the clock.* [PWGSC]

Figure 7.3 *Women did many different jobs on this project. They were Huisman technicians, security guards, labourers, nurses and engineers. They worked on surveying crews, tied rebar, and played a variety of roles in administration and communication.* [Boily]

Figure 7.4 *The cavernous space inside bridge girders is home to the steel tendons that tie the various bridge components together, and this space also provides a shore-to-shore corridor for utility cables. Island Tel already installed a fibre-optic communication cable through the corridor, and there is plenty of room for additional communication and electrical power cables.* [Boily]

Figure 7.5 *With many additional vessels now on the job, marine activity increased dramatically in the spring and summer of 1996.* [Boily]

Figure 7.6 *The* Svanen *prepares to load a pier base at nightfall. Marine work, like work in the fabrication yard, went around the clock during the 1996 construction season.* [PWGSC]

Figure 7.7 *The helicopter was an immense help during 1996. It was used to transport sandbags, pour concrete, transport personnel, and even carry portable toilets. Here, it carries concrete.* [Boily]

and the next thing you knew, every week another main girder came out the other end."

On the Water Early

With things in the Yard going well, the pressure was now on the marine crews who had to dredge the pier sites, place the components, and deliver grout and tremie concrete where it was needed. Rene Kolkman, the Marine Works Manager, talked about planning for this: "We wrote procedures and work methods for everything we would have to do, with everything specified down to the smallest detail. We identified the risks, and the areas where we could lose time. We coordinated the movement of every vessel, and it worked fabulously well."

It had been a mild winter, and the ice was expected to leave the Strait early. To take full advantage of that, marine crews and equipment were mobilized early, and the *Svanen* was able to place main girders on the 20th, 21st, and 22nd of April — about two weeks ahead of schedule. In addition, new vessels and new crews had been added to the fleet. Rene Kolkman again: "We were going to have to pour tremie concrete to seat the pier bases, and had to be able to supply at least 45 cubic metres per hour. This meant that we needed three supply boats. Besides that, we needed an additional dredger to prevent the dredging activity from slowing the project down. And we needed another crane barge to do the tremie concreting. We also needed better anchor-handling boats to handle the anchoring of the *Svanen*. With two boats we were able to do it in just one and a half hours every time.

Kolkman went on to say, "All the boats worked 7 days a week, 24 hours a day, with crews on for two weeks and off for one. The *Svanen* was completely crewed by Canadians — including Captains and everybody else. In 1995 we had brought in a

Figure 7.8 *In its busiest period Strait Crossing employed 42 marine vessels of various sizes.* [Boily]

Figure 7.9 *The* Svanen *picks up a main bridge girder* 192 *metres long and weighting* 7500 *metric tonnes. This is the longest and heaviest component that the* Svanen *lifted.* [*Boily*]

Figure 7.10 At 11:30 *p.m.* *on Tuesday, November 19, 1996 the* Svanen *placed the last structural component — the drop-in girder between pier 34 and pier 35. Fireworks erupted and workers cheered. The Confederation Bridge now spanned the Strait.* [Boily]

few Danish workers, but after training the Canadian crews these guys went home and didn't come back.

"Transporting as many as 500 or 600 people back and forth each day from marine areas was another big problem. We had two conventional boats that could carry about 40 people at a time at 10 or 12 knots maximum speed, but this was not good enough. So at the end of November 1995 we, together with some other people, started to design a boat. In the end we ordered two of them, each capable of carrying 16 people and traveling at 32 knots. They were very safe boats, with the bow designed to fit into the temporary platforms at the bridge piers. These custom-designed boats cost hundreds of thousands of dollars apiece. But based on the number of man-hours we were saving by getting the guys out there in 10 minutes flat, it was an easy calculation to show that buying them was the thing to do."

Early in 1996 a helicopter was brought in to expedite the construction process. It was used for transporting sand bags, pouring concrete, shuttling crews from place to place — even carrying portable toilets. The chopper was fast and adaptable, and proved especially helpful in the concrete work associated with installing Matchcast components.

The Busiest of Summers

As spring slipped into summer, the one remaining uncertainty in the fabrication yard was the production of drop-in girders. To remedy the situation, the process was refined to reduce cycle times, people were added, and production continued all night. Marine activities were also refined. "People were very inventive," recalls Rene Kolkman. "They came up with all kinds of gadgets and tricks to make things easier, better, faster, and still not lose quality. So when the busy July/August placement time came, we were ready. We placed 29 components in August."

Other Happenings

Progress toward physical completion was accompanied by progress on other bridge-related matters. One of these concerned an official name for the bridge. On May 31, Diane Marleau, Minister of Public Works and Government Services, announced the formation of an advisory committee to consider names for the new bridge. The public would submit suggestions, and the Committee, chaired by former P.E.I. Premier Alex Campbell, would review them. The committee considered 2200 entries, and as requested, submitted a short list of three possibilities to the Minister for her decision. On September 27 she announced that she had chosen *Confederation Bridge* — by far, the most commonly submitted name. This decision — like almost everything else connected with the project — stirred public controversy. There were media reports that the Committee had also expressed a favourite name of its own — one which differed from the Minister's. Islanders took sides on the issue, but Diane Marleau stuck by her decision, and *Confederation Bridge* it is.

In the area of compensation, ferry workers and Marine Atlantic reached agreement on severance arrangements, and Strait fishers would soon

Figure 7.11 *Only 24 main bridge components had been placed in 1995; the remaining 151 were placed between April and November of 1996. This amazing feat was something for all to celebrate, and the day after the last structural component was placed the company held a "topping off" party. Here, the Strait Crossing management team of (l. to r.) Gilles deMaublanc, Paul Giannelia, and Harmen Blom toast the success of the project — and the work force that made it happen.* [PWGSC]

receive their third and final year of compensation payments from Strait Crossing for disruption of fishing activities. No lobster fishing had been permitted in the construction corridor during the 3-year construction period, and this meant that in each of those years there were roughly 4000 fewer lobster traps in the water. Compensation was linked to the reduction of fishing effort. Lobster fishers who had normally put traps in the corridor had two choices. If they chose not to put those traps in the water at all, they received a flat fee plus $200 per trap per year. If they chose to put them somewhere else in the Strait, they received substantially less compensation — and other fishers were invited to increase their trap removal until the 4000 trap quota was met. Scallop, Herring, and Mackerel/groundfish fishers who historically used the corridor received $3000, $2000, and $1000 respectively. There was also compensation for Irish Moss and smelt fishers. Total compensation paid to all fishers during the three years approached $5 million.

Completing the Structure

Aided by excellent weather and the high-energy "let's get it done!" spirit of the Strait Crossing work force, phenomenal progress was made during the summer and fall. By early November there were only a few components left to be placed, and daily progress reports in the media allowed Islanders to follow the countdown toward completion. Finally, at 11:30 p.m. on Tuesday, November 19, 1996 the *Svanen* placed the last structural component — the drop-in girder between pier 34 and pier 35. Fireworks erupted and workers cheered. The Confederation Bridge now spanned the Strait. The fixed link was in place. The Island and mainland were connected. And being able to open the bridge June 1, 1997 — right on schedule — was now almost assured.

A Celebration, and a Winding Down

The main bridge comprised 175 large structural pieces: pier bases, pier shafts, main girders, and drop-in girders. Only 24 of them had been placed in 1995; the remaining 151 were placed between April and November of 1996. This amazing feat was something for all to celebrate, and the next afternoon the company held a "topping off" party. Attended by over 1000 employees, it was an occasion to celebrate a job well done. In his comments to the crowd, Paul Giannelia said, "Today is a day for professionals. Today is a day for champions. We are champions because we said we would do it, and we did it. We've got something out there in the Strait that we can be very proud of, and for the rest of our lives nobody can take it away from us."

Accomplishments are occasions for pride, but they also represent endings and times of transition. Kevin Pytyck reflected on this: "After chasing the goal for so long, and working so hard towards it, it's a different feeling when you've finally done it. The main part of the task is behind us now, and it will take a week or so for that to settle in. The sense of pride is overshadowing the realization that the project is rapidly coming to a conclusion."

Employment had been decreasing since the midsummer peak of about 2400, and would decrease even more rapidly now. Only about 150 employees continued to work over the winter. Many of the vessels moved on to other projects right away; others spent the winter in Georgetown, P.E.I. The *HLV Svanen* spent the winter there before heading across the Atlantic once again. As a member of the European Storebælt Group who had built the *Svanen*, Ballast Nedam had sold the vessel to Strait Crossing and was now buying it back — this time to build a bridge between Denmark and Sweden.

Figure 7.12 *An ironic moment on October 20, 1996, when the Marine Atlantic ferry* Abegweit *accepts the honour of being the first ship to use the bridge's navigation span.* [Boily]

A Legacy for Workers

About 60% of the craft work force — some 1200 to 1300 people — were members of Local 1077 of the Labourers' International Union. They were involved with rebar, concrete placement, concrete finishing, post-tensioning, security, materials handling, and several other activities. John P. Rose, the Local's Business Manager, reflected on what the project had meant to the membership: "Our members gained a lot from the project, especially from the training. At the time the bridge started we had on P.E.I. about 20 experienced rebar tiers. Later we had in excess of 500 up there tying rebar. They went in as apprentices and became journeymen."

Figure 7.13 *Workers on the Confederation Bridge project are dwarfed by the* Svanen. *[Boily]*

Sandy Clarke, the Local's President, added his comments: "Before the bridge, most of our people had only worked on commercial projects — nothing like the magnitude of this project. And working on this project opened their eyes to many things. The training they got is one thing they can take with them. Strait Crossing created a state-of-the-art rebar shop up in Slemon Park. Our people manned that shop for them, and the training and experience that those employees got in that shop was second to none. Just having worked on a megaproject is another major benefit. Things are done in a very different way — the coordination is much more complex — and they now know how they fit into the chain of command and how their work fits into the project. They successfully worked on a successful megaproject. The reputation of the project and their reputation within the project is something they can take to another job site.

"Some will choose to head off to other large projects — perhaps Voisey's Bay, the Sable Island pipeline, or the four-lane highway in New Brunswick. They now know how much they can make on these megaprojects, and some are willing to make the sacrifices they need to make in order to work on them. But whether they move on or stay here, this was a very positive project and experience. This was the biggest, and these people were helping to build it. There is this legacy. They can all point to the bridge and tell their kids, 'I helped build that.'"

ASSURING QUALITY

On the construction side, Strait Crossing contracted to build a bridge with a 100 year design life. On the operating side, they agreed to operate and maintain that bridge for 35 years, and then turn it over to the government in good enough condition to be used for another 65 years — and beyond.

This second part of the deal was a powerful incentive to build it right, and there were four different systems in place to ensure that this would happen:

Quality Control — Primary responsibility for quality existed at the worker and supervisor level — a personal commitment to do it right. This has become commonplace in the auto industry, but is fairly new in construction. On this job there were hundreds of written *Work Method* procedures — one for each fabrication and marine task. These work methods had to be adhered to, and it was the responsibility of both the workers and their supervisors to confirm that the specified procedures had been followed.

Quality Assurance — QA involved random checks in some situations, and regular testing in others. Every load of concrete, for example, was inspected in its plastic state for properties such as consistency (as measured by the "slump" test) and air entrainment. QA personnel also extracted concrete cores at random from the bridge, and assessed them for durability.

The Engineer of Record — Here, as in most infrastructure projects, the engineering firm responsible for the design was also responsible for ensuring that the design was properly executed. JMI/Stanley reviewed QC and QA records, and verified implementation.

The Independent Engineer — This bridge was designed by the developer, and not (as is usually done) by an engineering firm hired by government. Because of this, the federal government required the developer to have an "Independent Engineer" on the project. Buckland and Taylor is that firm. In addition to checking design compliance with project requirements, and monitoring the financial status of the project, Buckland and Taylor played a role with regard to quality. Their job was to monitor the inspection process and confirm for the government and the general public that quality standards were being met.

Was all this successful? Both Strait Crossing and the Independent Engineer feel that it was. Yes, mistakes were made. But when remedial steps were needed, they were taken. An expensive mistake occurred early in the project. Strait Crossing QA personnel found the tremie concrete seal under the E5 pier base (the first one installed) to be faulty.

As Kevin Pytyck tells the story: "It was a big problem and put us back several weeks. There were several hundred cubic metres of concrete that had to be removed to determine the extent of the problem, and in the end, 99.9% of it came up. Then we redid it. This was very early in the project, and it did two things. First, it demonstrated our commitment to quality. We're going to rip it apart, and it's going to cost the project half a million bucks. But it also taught a lot of our people how serious this quality business was. If you are going to have a problem, best it be the first time you do something. It was a hard lesson for everybody — management, supervision, and labour — and an important wake up call."

BRIDGE FACTS

Total volume of concrete placed on site:
478,000 cubic metres

Total tonnage of rebar placed on site:
58,500 tonnes

Total tonnage of post-tensioning cable:
13,960 tonnes

Total length of post-tensioning cable:
Approximately 12,690 km

Total volume of dredging:
277,100 cubic metres

Deepest pier location (Pier 26):
39.8 m below mean sea level

Total personnel at peak production:
2079 craft workers
415 staff

Total number of marine vessels at activity peak:
42

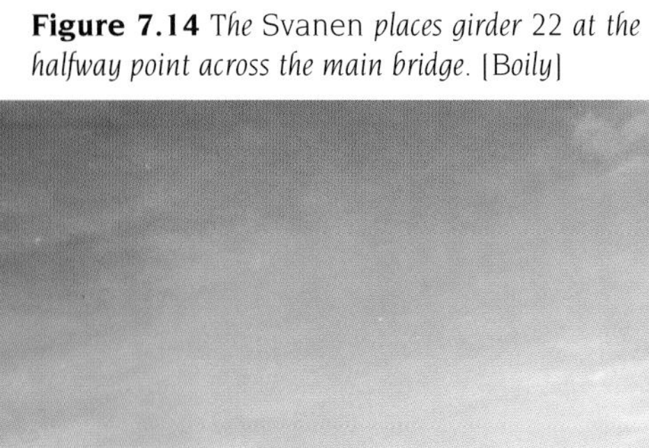

Figure 7.14 *The* Svanen *places girder 22 at the halfway point across the main bridge.* [Boily]

CHAPTER EIGHT

FINISHING TOUCHES

With the bridge structure in place, Strait Crossing's attention turned to bridge finishing work and bridge operation. Strait Crossing still had to

- turn the basic structure into a functional bridge by installing lighting, pavement, guardrails, etc., and
- complete the facilities and systems that Strait Crossing Bridge Ltd. required in order to operate the bridge safely and efficiently.

Bridge Features: A Focus on Safety

Bridges must be safe in two senses. The structure must be able to support the roadway and the users in a safe and reliable manner. And there must be operational safety. Drivers and their vehicles must be able to cross the bridge with minimal risk of an incident. The right structural features can contribute to operational safety, and the Confederation Bridge has several:

- **A spacious roadway.** At 11 metres overall width, the roadway not only has two 3.75 m traffic lanes, but two 1.75 m shoulders as well. These shoulders are wide enough to accommodate disabled autos.
- **1.1-metre-high safety barrier walls.** Built of steel-reinforced, high-strength concrete, these walls are intended to keep errant vehicles on the roadway and to minimize visual distraction. (See Figure 8.1)
- **A roadway with curves.** The three curves in the Confederation Bridge roadway help drivers stay alert by eliminating the hypnotic effect that a straight roadway sometimes induces.

An effective lighting system illuminates the roadway from one end of the bridge to the other, and operational safety is further enhanced by a high-tech *Traffic Management System* or *TMS*. Here, several electronic technologies have been integrated into one coordinated system. Designed and implemented by Strait Crossing and two other Canadian companies — Intrex Inc. and the IBI Group — the Traffic Management System uses sophisticated bridge monitoring to allow safety-conscious traffic control on a moment-to-moment basis.

The primary *inputs* to the TMS come from:

- **Emergency Callboxes.** Seventeen emergency callboxes located at 750 m intervals on the

Figure 8.1 *There is a 1.1-metre-high safety barrier wall on each side of the bridge to keep errant vehicles on the roadway and minimize visual distraction. Shown here under construction, the barrier is made of steel-reinforced high-strength concrete.* [PWGSC]

north side of the bridge put motorists in instant voice communication with bridge management personnel. Each callbox also has an emergency alarm button on the outside and a fire extinguisher inside.

- **Video Surveillance**. Seventeen video cameras are located at 750 m intervals on the south side of the bridge, and three others monitor the bridge approaches.
- **Weather Monitoring.** Wind velocity is monitored constantly, and in-pavement sensors detect conditions which warn of possible ice formation such as temperature, precipitation, and dew point.

All of the above feed into the *Bridge Management Centre*, a room located in the Bridge Operations Building at the P.E.I. end of the bridge. Callbox phone calls and alarms are answered there. TV camera outputs appear on a bank of monitors, and pan, tilt, and zoom control of each camera enables the operator to get a clear view of problems on the bridge. Weather data is presented to the operator in numerical form, and if wind velocity exceeds a preset level, an alarm sounds. Information about bridge conditions is continuously evaluated by Centre personnel who can respond to situations immediately with a wide range of computer-facilitated traffic control options. These include:

- **Standard Traffic Signals.** Seventeen individually-controlled red/amber/green signals in each direction allow traffic to be slowed, or stopped completely, according to circumstances.
- **Changeable Speed Limit Signs.** Placed at every other traffic signal, the speed limits which these signs display can be changed by the TMS operator whenever road and driving conditions dictate.

Figure 8.2 *Once the bridge was complete workers finished the surface, lighting and other features.* [PWGSC]

Figure 8.3 *Ice monitoring is part of the ongoing research effort. The group shown here is about to lower an upward-looking sonar device to the bottom of the Strait to see if it can measure the keel depth of floating ice.* [PWGSC]

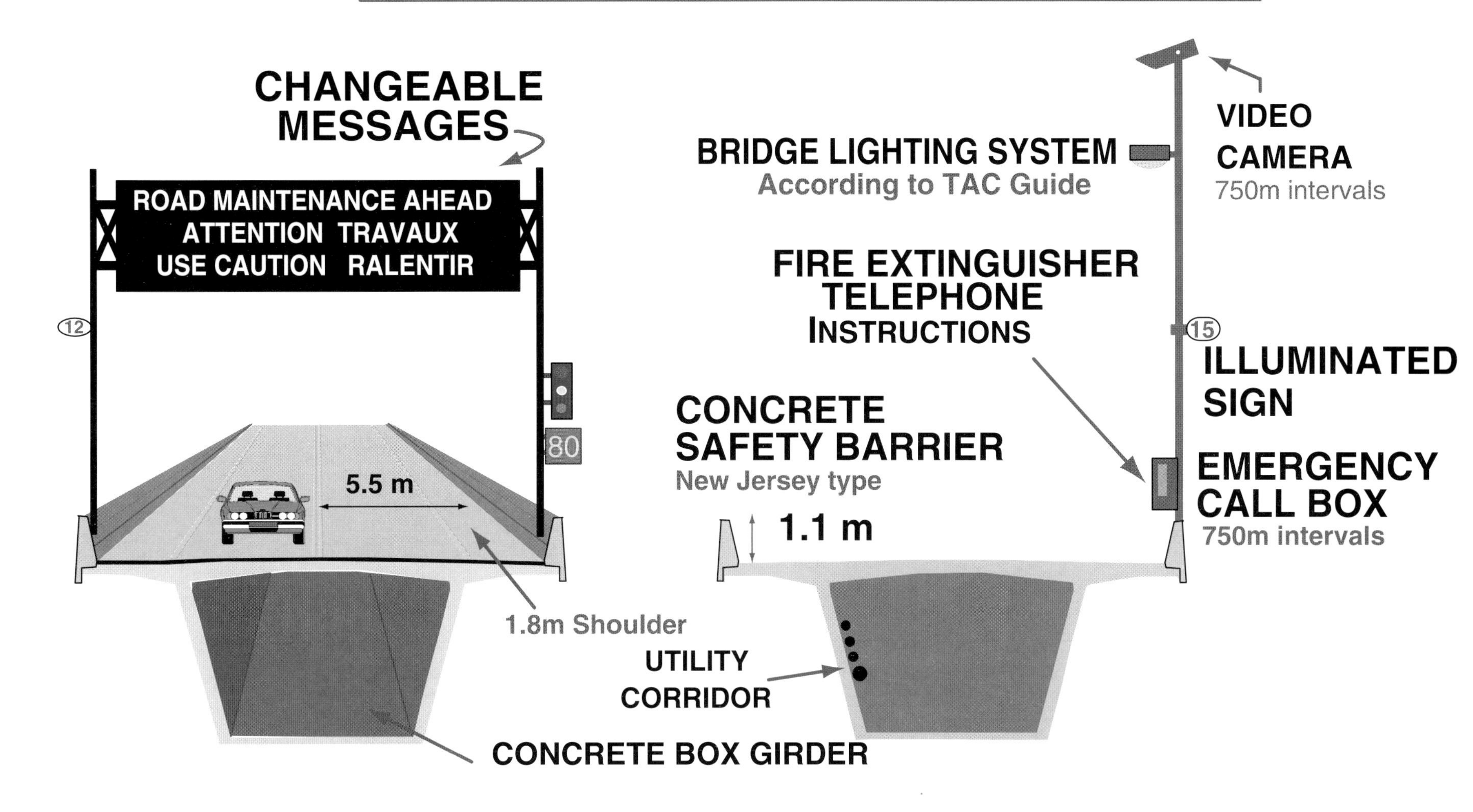

Figure 8.4 *An effective lighting system illuminates the roadway from one end of the bridge to the other, and operational safety is further enhanced by a high-tech* Traffic Management System, *involving several electronic technologies integrated into one coordinated system.* [PWGSC]

- **Changeable Message Signs.** Located at the bridge entrances on both sides, these large signs display messages about roadway conditions to those who are about to cross the bridge.

Some of these features are shown in Figure 8.4 (see previous page).

A round-trip bridge toll is collected on the P.E.I. side of the bridge from travellers leaving the Island. At the P.E.I. toll plaza there are seven toll lanes — staffed according to need, and accepting cash, credit cards, and debit cards. A self-serve lane for cars takes credit and debit cards only. Foot passengers and cyclists pay their tolls in the waiting room.

At the normal travelling speed of 80 km/hr, the trip across takes about 12 minutes. In general, any vehicle permitted on the Trans Canada Highway is permitted on the bridge. However, under exceptionally windy or stormy conditions it may occasionally be necessary to restrict the access of certain types of vehicles such as motorcycles, high-sided vehicles, and empty trucks. Also, special controls apply to the movement of oversized and slow-moving vehicles, and vehicles carrying hazardous materials. Depending on the goods classification and crossing conditions, the Bridge Patrol might escort the vehicle across or require that the crossing be delayed. Pedestrians and cyclists are not allowed on the bridge, but are taken across by a shuttle bus that makes frequent runs.

Completing Construction

During the winter of 1996/97, work continued in several areas. On the bridge itself, workers continued to install expansion joints where drop-in girders joined main girders, and they installed and wired lighting fixtures. Then, when the weather warmed sufficiently, work resumed on casting the last 4.5 km of safety barrier, and bridge paving got under way.

The paving process resembled roofing work as much as it did road paving. A layer of rubberized asphalt was applied to the concrete surface of the bridge. This created a smooth surface upon which a waterproof membrane was placed. The roadway driving surface — a layer of aggregate with an asphaltic binder — was then applied to the membrane. This 3-layer approach gives the bridge considerable protection against salt and other road chemicals. It's a high-durability system intended to minimize roadway and bridge maintenance.

Work also continued through the winter on the Bridge Operations Building, the building that would house the cyclist and pedestrian waiting room, the Traffic Management System, toll reconciliation, and other activities related to day-to-day bridge operation. Installing, testing and debugging the Traffic Management System continued through the spring.

The Bridge Operations Building is located in Borden-Carleton's new Gateway Village complex, and other Gateway Village structures were also under construction during the winter and spring. The goal was to get them ready for bridge opening on June 1 and the BridgeFest celebration that led up to opening. One of the bridge exits leads right into Gateway Village, making it a possible "first stop" for visitors to the Island. Among the initial Gateway Village offerings are interpretive displays concerning Island heritage and culture, and tourist information.

Links with Community

When Strait Crossing got the contract to build and operate the Northumberland Strait bridge, it was clear that the company would be part of Island life for a long time to come. What kind of neighbour would Strait Crossing be?

Figure 8.5 *A view of the bridge from the water.* [PWGSC]

One neighbourly act was Strait Crossing's involvement with another Island bridge. In early discussions with the P.E.I. government, then Premier Joe Ghiz made it clear to Paul Giannelia that while Islanders might differ in their feelings about a bridge across the Strait, Islanders en masse would be eternally grateful to whoever managed to replace that traffic bottleneck connecting Charlottetown and Stratford: the Hillsborough Bridge. Money that Strait Crossing saved on some Confederation Bridge highway work was used to fund the engineering effort. Dr. Gamil Tadros of SCI and his associates addressed the problem and came up with an innovative bridge expansion and road improvement design that would cost much less than replacing the existing bridge. Strait Crossing Inc. and the Province reached agreement on price and financing. Work began in August 1996, and is scheduled to be completed in the fall of 1998. Construction is being phased in over two years to minimize interference with the 25,000 cars that cross that bridge each day.

Another neighbourly act was Strait Crossing's response to Islander interest in Confederation Bridge construction activities. "I've never been on a project where everyone was so interested in how you do it," said Kevin Pytyck. "The interest among school children was unbelievable, and some of the things we did were education related. In partnership with one of the newspapers — the *Guardian* — we developed a teacher's resource guide. Also, myself, the staff, and the engineers did hundreds of in-class 'show and tells' about the bridge. And it was not just bridge structure that these students were interested in. At the high school and university level they wanted to know all about our environmental policies, strategies, and management. Responding to this interest was a major project for us, but it would have been a missed opportunity if these young people hadn't had a chance to learn about the project while it was going on."

Adults in large numbers were also interested in the project. Here, Strait Crossing responded in a couple of ways. They opened an information office at the corner of Water and Queen Streets, in the heart of downtown Charlottetown, and gave out thousands of information kits to people who stopped by. Arranging public tours of the Borden-Carleton Staging Facility was another response. "The phone was ringing off the wall the first year we were here, but we had nothing to show people," recalls Kevin Pytyck. "Then in 1995 we put a tour package out to tender. If we could be cost neutral on our own costs (printing literature, making models, etc.) and a tour operator could make some money on it as a business, great, then everybody is happy. The tour operator has a business going, we've looked after a big demand that the community has, and people get to see the project. In 1995 we had 40 to 45 thousand people take these tours, plus thousands of people on special tours — engineering associations, and so forth. In 1996 there were 75 or 80 thousand. So over the course of the project, over 125,000 people took that public tour. Many people got deeply interested in the technical details, and we now have bridge experts all over the Island. Some build model bridges in their basements; others have come up with quite profound questions."

Kevin Pytyck went on to say, "All this was simply where our basic attitude took us. We felt that our success in getting the contract had a lot to do with being straightforward, accessible, transparent — and it wouldn't make any sense to change into a closed corporation the day we signed the contract. If you're honest and open with the public, nothing is going to sneak up on you. So we took the approach that you had to be accessible and you had to be

Figure 8.6 *The bridge in November 1996.* [PWGSC]

transparent, and that's why we responded to all those school groups, and responded to the desire for tours, and why we have a Web page."

Today, many of the organizations that add richness and diversity to Island life depend largely — or solely — on contributions from individuals and corporations to carry on their activities. Increasingly, corporations are recognizing this and are choosing to help out. From 1993 through 1996 Strait Crossing supported over 300 community, cultural, recreational, and charitable organizations and activities through donations of time, money and merchandise. This outreach included a 5-year sponsorship of the Confederation Centre of the Arts production of the *Anne of Green Gables* musical, and substantial contributions to Summerside's College of Piping, the IWK Children's Hospital telethon, and Gold Cup and Saucer Parade prizes and broadcasts. Of less monetary value, but equally important to the organizations involved, were numerous sponsorships of sports teams and events, musical events, and theatre groups. Strait Crossing workers contributed too. For instance, an on-site bottle drive, sponsored jointly by the workers and the corporation, raised tens of thousands of dollars. By early 1997 the total value of all these contributions exceeded half a million dollars.

The Research Lab in the Strait

Yet another facet of this amazing project is the Confederation Bridge's role in furthering environmental and engineering knowledge. As an integral part of the project, Strait Crossing agreed to fund an extensive Environmental Effects Monitoring Program where scientists study the effect of the bridge on about a dozen important aspects of the environment. These include ice climate, ice scouring of the seabed, water chemistry, shoreline erosion, benthic (bottom dwelling) habitat and communities, phytoplankton production, and seabird migration. Baseline or "pre-bridge" data was taken in all areas of interest during 1993 and 1994, and data collection continued during the construction years of 1995 and 1996. The extent of future monitoring will depend on how quickly and thoroughly collected data answers the various environmental questions that have been posed — and some long-term monitoring is anticipated.

The other area of research concerns engineering aspects of the bridge itself, and the effect of environmental factors such as wind, ice, and earth movement on the bridge structure. Strait Crossing, several universities, and Public Works and Government Services Canada are jointly funding a monitoring and research program encompassing six major areas of investigation:

1. **ice forces**, and relating those forces to Strait ice conditions;
2. **traffic loads** and their effect on the bridge;
3. the **dynamic response of the bridge** to transient wind, ice impact, seismic, and traffic loads;
4. **temperature and thermal effects**;
5. **long and short term deformation** of structural elements; and
6. **corrosion** of the reinforcing steel.

This program will allow design assumptions to be checked against measured reality. Some of the knowledge gained may be of direct use to Strait Crossing in guiding its bridge maintenance activities in the coming years. Other knowledge will facilitate the design of future structures. At present, there are no guidelines or codes covering the design of long-span bridges, so this research will contribute significantly to the art and science of bridge engineering.

Figure 8.7 *The Confederation Bridge looking from Borden-Carleton, Prince Edward Island towards New Brunswick.* [PWGSC]

The Island and the New Reality

The Confederation Bridge is a new *physical* reality that several thousand people brought into being through their imagination and inspired effort. It is also a new element in a constantly changing *societal* reality. Prince Edward Islanders value social cohesion, natural beauty, and caring ways — and don't want to lose those things. During this century controversies arose over several perceived threats to those values: the automobile, consolidated schools, and a bridge to the mainland. The bridge, like the automobile and the consolidated school, will bring change to the fabric of Island life, but like these other changes, its effect on P.E.I. society won't be known for some time.

Some predictions about the effect of the bridge will probably come to pass. Some won't. And some things will happen that no one predicted. With ferry jobs gone, will Borden-Carleton become a ghost town? Or will it become a commercial centre, an industrial centre, or even an artistic centre? Only time — and imagination and effort — will tell. With the bridge in place, will people come to the Island in droves and buy up all the prime land? Only time — and what Islanders decide to do about land ownership and use — will tell.

The future of Island society did not depend entirely on whether automobiles replaced horses and carriages or whether consolidated schools replaced one-room schools. These were highly controversial issues in their day. But when the changes came, Islanders accepted them and moved on — while continuing to value neighborliness, community, and the beauty of their Island. Something similar will likely happen with the bridge, and this controversy, too, will fade away. Time will tell.

About some things, there is no controversy at all. Debt-strapped governments can no longer handle public infrastructure projects in the old way. Private-sector involvement is here to stay, and with it will come some of the things pioneered on this project:

- Smarter, more cost-effective designs — resulting in part from tight integration of structure designing and construction planning.
- Making environmental protection, management, and monitoring an integral, high-ranking part of the job.
- Analyzing socio-economic impacts early, and including appropriate economic mitigation in the overall plan.
- Private sector financing — so that governments can have these projects without going further into debt.
- The one-stop shopping approach: Design/Build/Operate/Transfer/Finance — or some combination of those elements.

From a design and construction perspective, this project went beyond ordinary again and again. There were the bridge's gravity-based foundations, its long spans, its 100 year design life, its high safety factor, the incredible construction pace during 1996, and its use of the largest bridge components ever built on land and then placed by vessels. In March 1997, the Canadian Construction Association gave Strait Crossing Inc. its Montgomery Memorial Award "for outstanding innovation in construction practices" — referring especially to the way Strait Crossing solved construction problems, its use of GPS with 2 cm accuracy to place components, and its use of the 100 tonne Matchcast Template to align the 7500 tonne main girder.

Without doubt, the Confederation Bridge is an amazing feat of engineering, construction, personal vision, and dedicated effort — and a personal triumph for all those who helped make it happen. Pride wells up, even in those of us who watched it happen from the sidelines.

BIBLIOGRAPHY

A great deal of information about the Confederation Bridge project is available on the Internet at Strait Crossing's Web site:

http://www.peinet.pe.ca/SCDI/bridge.html

Chapter 1 Project and Place

Atlantic Canada in the Global Community, Scarborough, Ontario: Prentice Hall Ginn Canada, 1997.

Douglas Baldwin, *Land of the Red Soil: A Popular History of Prince Edward Island*, Charlottetown, P.E.I.: Ragweed Press, 1990.

F.W.P. Bolger, Editor, *Canada's Smallest Province: A History of Prince Edward Island*, Charlottetown, P.E.I.: The Prince Edward Island 1973 Centennial Commission, 1973.

Deirdre Kessler, *Discover Canada: Prince Edward Island*, Toronto: Grolier Limited, 1996.

George Peabody, et al., Editor, *The Maritimes: Tradition, Challenge and Change*, Halifax, Nova Scotia: Maritext Limited, 1987

Chapter 2 Boats and Ice

Harry Bruce, "Chapter 2 — Across the Northumberland Strait," in *Lifeline: The Story of the Atlantic Ferries and Coastal Boats*, Toronto: Macmillan of Canada, 1977, pp. 100-167.

Lorne C. Callbeck, "Sagas of the Strait," *Atlantic Advocate*, Vol. 49, (February 1959)

Silver Donald Cameron, "The Ferry Services of Prince Edward Island," in *Iceboats to Superferries: An Illustrated History of Marine Atlantic*, St. John's, Newfoundland: Breakwater Press, 1992, pp. 61-99

Mary K. Cullen, "The Transportation Issue: 1873-1973," in F.W.P. Bolger, Editor, *Canada's Smallest Province: A History of Prince Edward Island*, Charlottetown, P.E.I.: The Prince Edward Island 1973 Centennial Commission, 1973.

Frank MacKinnon, "Communications Between P.E.I. and the Mainland," *Dalhousie Review*, vol. 29, (July, 1949), pp. 182-190.

Chapter 3 Exploring Alternatives

Harry Baglole and David Weale, *Cornelius Howatt: Superstar!*, Charlottetown, P.E.I.: H. Baglole and D. Weale, 1974.

Boyde Beck, "Tunnel Vision," *The Island Magazine*, No. 19, (Spring/Summer 1986), pp. 3-8.

Lorraine Begley, *Crossing that Bridge: A Critical Look at the P.E.I. Fixed Link*, Charlottetown, P.E.I.: Ragweed Press, 1993.

Public Works Canada, *CALL FOR EXPRESSION OF INTEREST, Northumberland Strait Crossing, Jourimain Island, New Brunswick to Borden Point, Prince Edward Island, Stage I*, May 12, 1987.

Institute of Island Studies, *Public Meetings on the Fixed Crossing: A report to the Office of the Premier, December 1987 to January 1988*, Ottawa: Public Works Canada.

P. Lane and Associates Ltd. / Washburn and Gillis Associates Ltd, *GENERIC INITIAL ENVIRONMENTAL EVALUATION of the Northumberland Strait Crossing Project*, Ottawa: Public Works Canada, March 15, 1988.

Public Works Canada, *Northumberland Strait Crossing Project: Proposal Call*, Ottawa: Public Works Canada, March 15, 1988.

Chapter 4 Evolution of a Solution

G. Barry, et al., *Final Report of the Ice Committee*, Ottawa: Public Works Canada, December 20, 1991.

G. Barry, et al., *Addendum to Ice Committee Final Report: Analysis of the Impact of SCI's Bridge Design on the Ice Environment*, Ottawa: Public Works Canada, April 1993.

Delcan – Stone & Webster, *Northumberland Strait Crossing Project: Bridge Concept Assessment*, Ottawa: Public Works Canada, April 25, 1989.

Delcan – Stone & Webster, *Northumberland Strait Crossing Project: Supplement to the Bridge Concept Assessment*, Ottawa: Public Works Canada, December 15, 1989.

FEARO Panel, *The Northumberland Strait Crossing Project: Report of the Environmental Assessment Panel*, Ottawa: Federal Environmental Assessment Review Office, 1990.

Jacques Whitford Environment, *Environmental Evaluation of SCI's Proposed Northumberland Crossing Project*, Fredericton, N.B.: Jacques Whitford Environment Limited, April 22, 1993.

Northumberland Strait Crossing Project: Federal-Provincial Agreement, Ottawa: Public Works Canada, December 16, 1992

Northumberland Strait Crossing Project, *STRAIT FACTS* (newsletter), Charlottetown, P.E.I.: Public Works Canada, Vols. 1-19, June 1987 to Winter 1995-96.

Public Works Canada, *Northumberland Strait Crossing Project: Federal Government's Response to the Recommendations from the Report of the Environmental Assessment Panel*, Ottawa: Public Works Canada, November 21, 1990.

Public Works Canada, *Northumberland Strait Crossing Project: Stage II Proposal Call Consolidated*, Ottawa: Public Works Canada, June 3, 1991.

Public Works Canada, *Northumberland Strait Crossing Project: Consolidated Stage III Proposal Call* Ottawa: Public Works Canada, May 20, 1992.

Strait Crossing Inc., *Environmental Management Plan for the Northumberland Strait Crossing Project*, Calgary: Strait Crossing Inc., February 15, 1993.

Chapter 5 Hitting the Ground Running! — 1993/94

William M. Ainley, "Making B-O-T Work for a Major Bridge Project," *Public Works Financing*, November 1993.

William G. Reinhardt, "P.E.I. Bridge Project Essentials," *Public Works Financing*, November 1993.

Chapter 6 A Year of Milestones — 1995

Barry Lester, P.Eng., Gamil Tadros, Ph.D. P.Eng., "Northumberland Strait Crossing: Design Development of Precast Prestressed Bridge Structure," *PCI Journal*, September-October 1995, pp. 32-44.

George A. Peer, "Getting Ready to Erect the Big Ones," *Heavy Construction News*, January 1995.

George A. Peer, "P.E.I. Bridge Activity Gains Momentum," *Heavy Construction News*, July 1995.

George A. Peer, "NS Stone Goes Into P.E.I. Bridge Segments," and "Strait Piers Rise Near N.B. Shore," *Heavy Construction News*, August 1995.

Chapter 7 Toward Completion — 1996

Amin Ghali, Gamil Tadros, and Paul H. Langohr, "Northumberland Strait Bridge: Analysis Techniques and Results," *Canadian Journal of Civil Engineering*, Vol. 23: 86-97 (1996).

Peter Green, "Placing Spans at a Dire Strait," Engineering News-Record, September 16, 1996.

Wilbert S. Langley, "Northumberland Strait Crossing Project: Materials, Specifications, and Construction Techniques," Dartmouth, NS: Jacques, Whitford and Associates, 1996.

George A. Peer, "P.E.I. Bridge Gets Its Feet Wet," *Heavy Construction News*, January 1996.

Gerard Sauvageot, J. Muller International, "Northumberland Strait Crossing, Canada," Fourth International Bridge Engineering Conference.

Strait Crossing and Public Works & Government Services Canada, *ENVIRONMENT: The Northumberland Strait Bridge Project*, Charlottetown: Strait Crossing and Public Works & Government Services Canada, 1996, 23 pages.

Chapter 8 Finishing Touches — 1997

Ron Allen, et al., "Driving to Prince Edward Island: Traffic Management Issues on the Northumberland Strait Bridge," Toronto: IBI Group, July 1996.

Ross Gilmour et al., "Snaking Across the Strait: Northumberland's Ice Breaker," *Civil Engineering*, January 1997.

ILLUSTRATION CREDITS

[Boily]	Boily Photo, Summerside, P.E.I.
[JMI]	J. Muller International, San Diego, CA
[Journal-Pioneer]	Journal-Pioneer, Summerside, P.E.I.
[Marine Atlantic]	Marine Atlantic, Moncton, N.B.
[Masterfile]	Masterfile Corporation, Toronto, Ontario
[PAPEI]	Prince Edward Island Public Archives and Records Office
[PEIM&HF]	P.E.I. Museum and Heritage Foundation, Charlottetown, P.E.I.
[PWGSC]	Public Works and Government Services Canada
[SCDI]	Strait Crossing Development Inc.

Special thanks to Wout Krispijn for endpaper drawing.

Construction Confederation Bridge Sept. 1996